Color Atlas of
Xenopus laevis Histology

Color Atlas of
Xenopus laevis Histology

by
ALLAN F. WIECHMANN, PH.D.
Departments of Cell Biology and Ophthalmology
University of Oklahoma

and

CELESTE R. WIRSIG-WIECHMANN, PH.D.
Departments of Cell Biology and Zoology
University of Oklahoma

Springer Science+Business Media, LLC

Library of Congress Cataloging-in-Publication Data

Wiechmann, Allan F., 1954-
    Color atlas of Xenopus laevis histology / by Allan F. Wiechmann and Celeste R. Wirsig-Wiechmann.
        p. cm.
    Includes bibliographical references and index.
    ISBN 978-1-4613-4876-4   ISBN 978-1-4419-9286-4 (eBook)
    DOI 10.1007/978-1-4419-9286-4
        1. Xenopus laevis—Histology—Atlases. I. Wirsig-Wiechmann, Celeste R., 1955- II.
    Title.
QL668.E265 W54 2003
597.8'654—dc21   2002041377

Copyright © 2003 Springer Science+Business Media New York
Originally published by Kluwer Academic Publishers in 2003
Softcover reprint of the hardcover 1st edition 2003

Permission for books published in Europe: permissions@wkap.nl
Permissions for books published in the United States of America: permissions@wkap.com

Printed on acid-free paper.

To Paul, Elliot and Nicholas
Our most fruitful collaborations

Preface:

The stimulus to write a color atlas of *Xenopus laevis* histology arose from the needs of the author's research program. We had developed antibodies specific for some receptors in the African Clawed Frog, *X. laevis*, and although our major interest was the retina, we were also interested in examining the receptor distribution in many organs throughout the organism. It was then that the significant differences between amphibian histological structures and mammalian histology became apparent to us.

Our search for a central source that described the microscopic anatomy of cells and organs of the adult *X. laevis* or any other amphibian was not successful. We found this somewhat surprising, given that *X. laevis* is an important model for early development, intercellular and intracellular signaling, cell surface receptors, chromosome replication; cytoskeletal and nuclear assembly; cell cycle progression, and vision physiology. Also, recent advancements in transgenic *X. laevis* technology make this an ideal model for the study of expression of genes and their functions. To facilitate the interpretation of the anticipated acceleration of new transgenic *X. laevis* studies, a central source of *X. laevis* histology was needed.

As the directors of the Medical Histology, Dental General Histology and Oral Histology courses at the University of Oklahoma Health Sciences Center, combined with our research experiences with amphibian models, we felt that we were poised to produce a central source of *X. laevis* histology images. The purpose of this atlas is to serve as a guide to assist scientists to gain a better understanding of the appearance and distribution of cells and other components that comprise the organs and tissues of *X. laevis*.

The images presented in this atlas were captured on an Olympus stereo photomicroscope equipped with a Spot camera. Some low magnification images were obtained on a Nikon dissecting stereomicroscope equipped with a digital camera. The tissues examined were obtained from male and female adult *X. laevis* that were perfused with a 4% formaldehyde, 2% glutaradehyde fixative. The organs were processed for paraffin sectioning, and then stained with hematoxylin and eosin (H&E).

We greatly appreciate the hard work and excellent technical assistance of Ms. Carla Hansens, the Histology Technician in the Department of Cell Biology at the University of Oklahoma Health Sciences Center. She embedded, sectioned, mounted and stained the specimens for the over 1,000 slides used in this project. Without her admirable work ethic and devotion to her craft, this atlas would never have become a reality. We also thank our Cell Biology Department chairman, Dr. Robert E. Anderson, who contributed greatly to this work by supporting our access to the technical assistance and facilities needed to create these images. We thank Rajendra Dohte and Radhika Dighe for their valuable help in assembling the index.

Allan F. Wiechmann, Ph.D.
Celeste R. Wirsig-Wiechman, Ph.D.

Contents

Chapter 1

Basic Tissues

Body organs are composed of a combination of four basic tissue types, although one type is usually predominant. The four basic tissue types are epithelial tissue, connective tissue, muscular tissue, and nervous tissue. Each of the basic tissue types will be briefly presented in this chapter, but will also appear in figures throughout this atlas.

I EPITHELIAL TISSUE

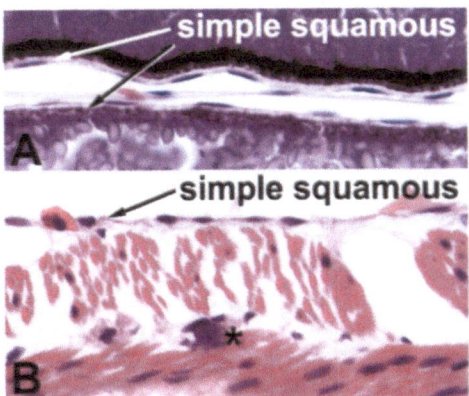

Figure 1. Simple squamous epithelium consists of a single layer of flattened cells that typically line a lumen, body cavity, or surface. **A.** Simple squamous epithelium lining the outer surface of the oocytes. **B.** Simple squamous epithelium lining the outer surface of the colon. The epithelium and underlying connective tissue make up the serosa. Directly underlying the serosa is smooth muscle in cross-section. At the lowest part of the figure is smooth muscle in longitudinal or oblique orientation. Between the two muscle layers is a neuron (*), which is part of the post-ganglionic parasympathetic plexus called Auerbach's plexus, or the myenteric plexus, which innervates the smooth muscle cells.

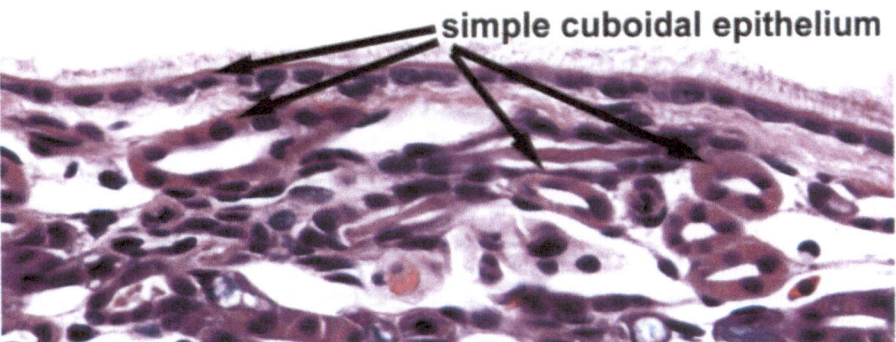

Figure 2. Simple cuboidal epithelium. The simple cuboidal epithelium lining the external surface of the kidney is ciliated, and the distal convoluted tubules are lined by simple cuboidal epithelium (arrows). The height and width of cuboidal epithelium is approximately equal, resulting in a cube-shaped cell.

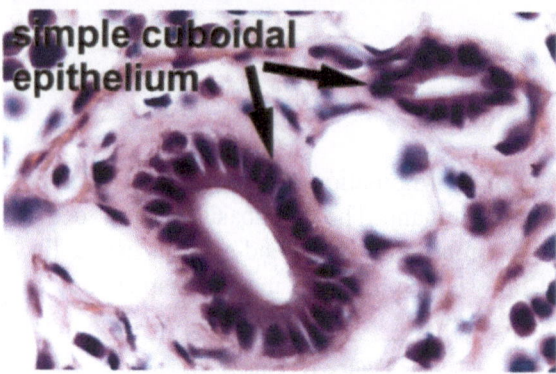

Figure 3. Simple cuboidal epithelium. Simple cuboidal epithelial cells line the inner surface of ducts in the pancreas. The epithelium of the larger duct could also be classified as low columnar epithelium. Loose connective tissue surrounds the ducts.

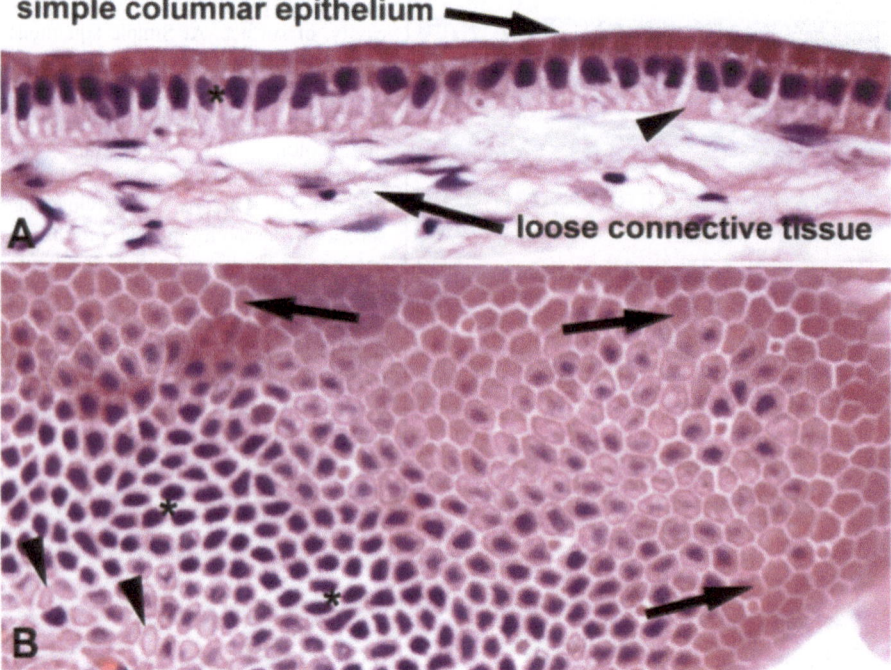

Figure 4. Simple columnar epithelium. **A.** Simple columnar epithelium lines the lumen of the gall bladder. The columnar cells are taller than they are wide. The epithelium is observed in a longitudinal orientation. Loose connective tissue underlies the epithelium, as it does in all organs. **B.** Simple columnar epithelial cells of the gall bladder shown in an oblique section. Note the regional differences in morphology from one region of the cell to the next. The most apical portions of the cells demonstrate a hexagonal shape and the cytoplasm is eosinophilic as indicated by the pink staining (arrows). In the nuclear region, there is only a thin rim of cytoplasm surrounding the basophilic (blue) nuclei (indicated by asterisks in A and B). The basal portion of the epithelial cells are not as intensely stained as in the apical portion (indicated by arrowheads in A and B), and are in contact with a thin basement membrane.

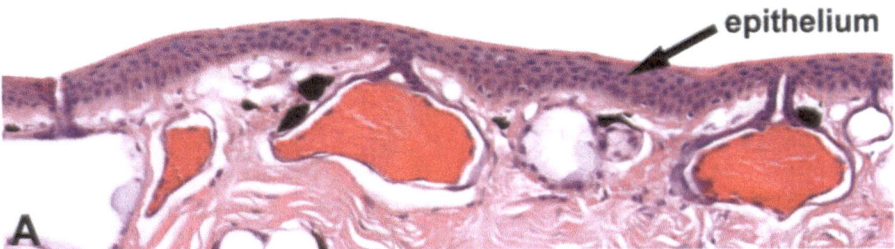

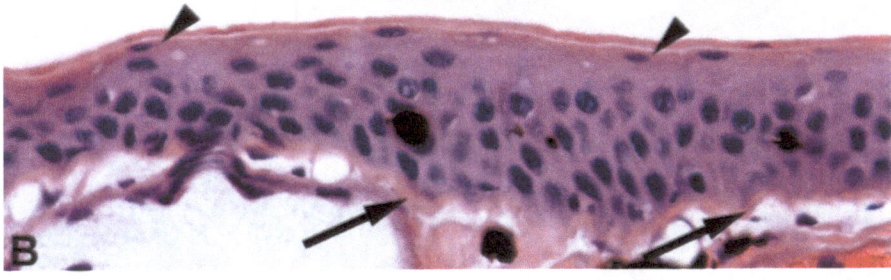

Figure 5. Stratified squamous keratinized epithelium. **A.** Low magnification of the stratified squamous keratinized epithelium (arrow) of the epidermis of the skin. Note the presence of exocrine glands and dense irregular connective tissue in the underlying dermis. **B.** High magnification of the epidermis. Arrowheads indicate the superficial layer of squamous epithelium. There is a thin layer of protective keratin above the cells. Note the thick basement membrane (pink) indicated by arrows on the basal surface of the epithelium which separates the epithelium from the underlying loose connective tissue. The dark brown cells that appear in the epidermis are pigmented cells.

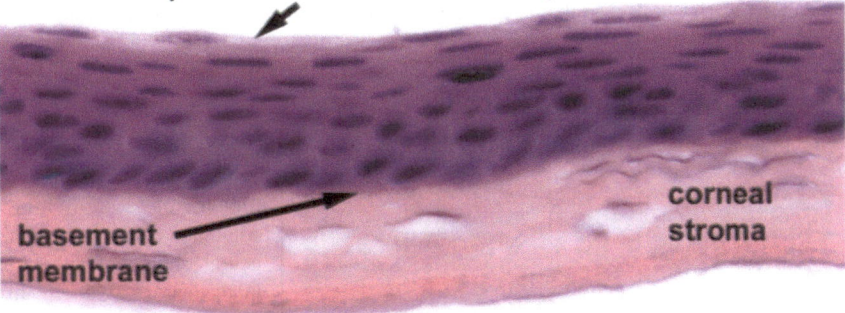

Figure 6. Stratified squamous non-keratinized epithelium. **A.** The simple stratified non-keratinized epithelium (arrow) of the epidermis of the cornea. The short arrow indicates the superficial layer of squamous epithelium. Note the basement membrane indicated by athe long arrow on the basal surface of the epithelium which separates the epithelium from the underlying connective tissue.

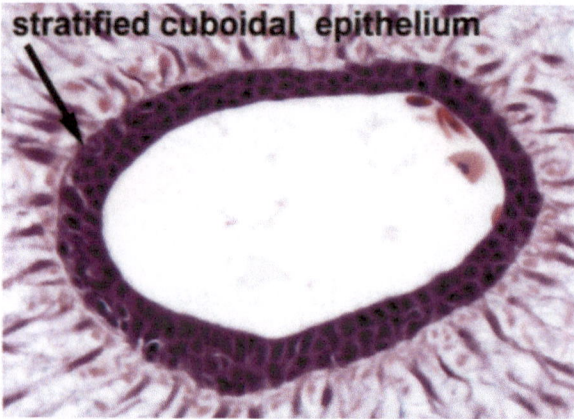

stratified cuboidal epithelium

Figure 7. Stratified cuboidal epithelium. Stratified cuboidal epithelium is comprised of two or more layers and the superficial layer of cells are cuboidal in shape. This type of epithelium often lines large ducts and other passages.

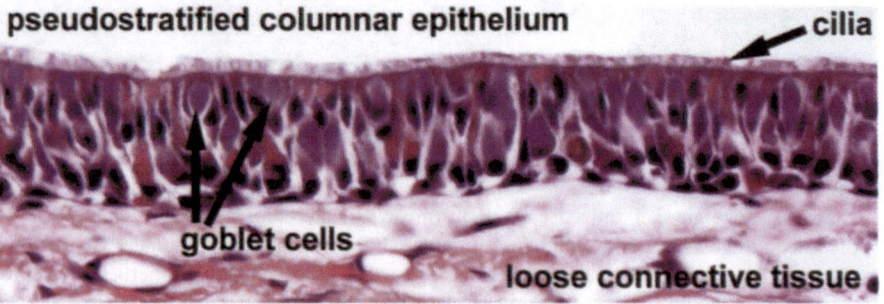

pseudostratified columnar epithelium   **cilia**

goblet cells

loose connective tissue

Figure 8. Pseudostratified columnar epithelium. Pseudostratified ciliated columnar epithelium with goblet cells lines the airway of the trachea, and is often referred to as respiratory epithelium. Note that in pseudostratified columnar epithelium, all of the epithelial cells contact the basement membrane, but not all of them reach the apical surface. There is a population of basal cells in contact with the basement membrane that give rise to new epithelial cells that will later contact with the apical surface.

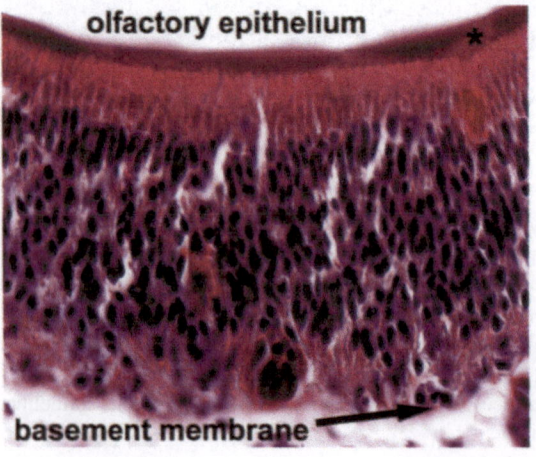

olfactory epithelium *

basement membrane

Figure 9. Pseudostratified columnar epithelium. An example of pseudostratified columnar epithelium is the neural olfactory epithelium in the nasal cavity. In this example, the sensory cilia on the apical surface are obscured by a layer of mucous (*). The columnar epithelial cells all contact a basement membrane.

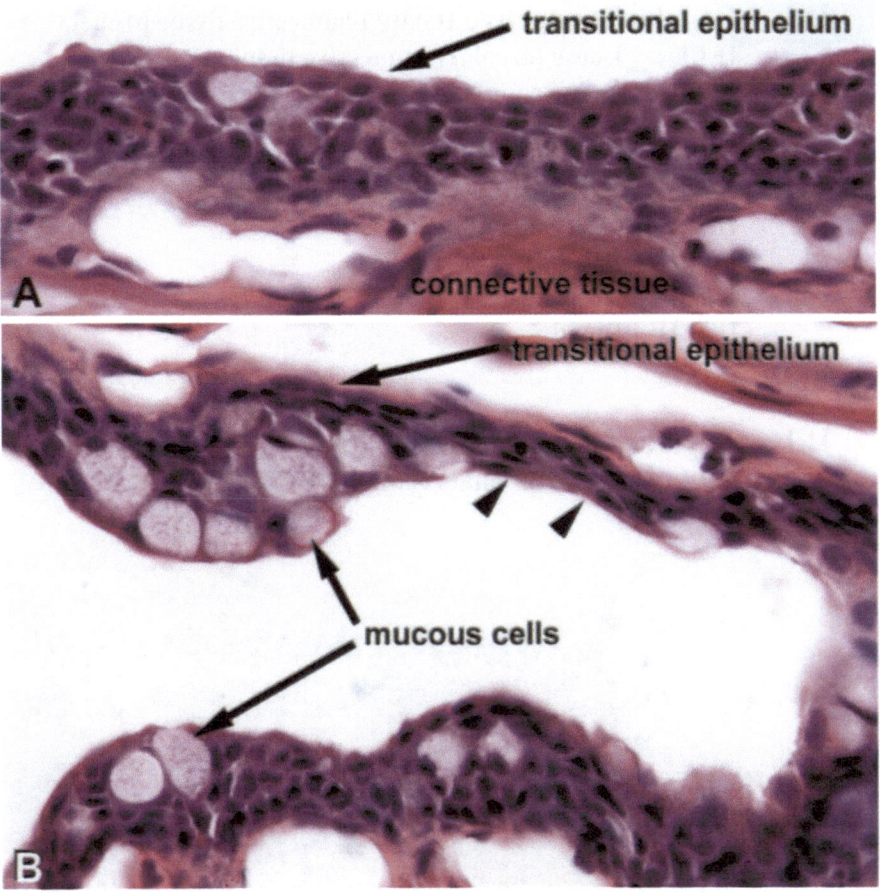

Figure 10. Transitional epithelium. **A.** Transitional epithelium lining the lumen of the urinary bladder. It is a stratified epithelium, with rounded (or "dome-shaped") cells on the apical surface. **B.** Transitional epithelium of the urinary bladder demonstrating the presence of goblet cells whose apical membranes are in contact with the lumen, so mucus can be secreted into the lumen. The major characteristic of transitional epithelium is its ability to modify its shape in response to distention. The epithelium in A and most of B is in the relaxed or non-distended configuration. When the bladder fills, it stretches the epithelium, which results in a squamous appearance in most of the epithelial cells, especially the surface cells (arrowheads).

II CONNECTIVE TISSUE

Connective tissues are classified according to their basic functions, which is to provide structural and metabolic support for the tissues of the body. They are subdivided into regular connective tissue (CT), and specialized CT, which include cartilage, bone, adipose and blood.

Classification of connective tissues

II.I   Regular connective tissue (connective tissue proper)

    II.I.I.   **Loose (areolar) connective tissue**

    II.I.II   **Dense connective tissue**

        A.   Dense irregular connective tissue

        B.   Dense regular connective tissue

II.II   Specialized connective tissues

    II.II.I   **Cartilage**

    II.II.II   **Bone**

    II.II.III   **Adipose**

    II.II.IV   **Blood**

II.I   Regular connective tissue (connective tissue proper)

    II.I.I.   **Loose (areolar) connective tissue**

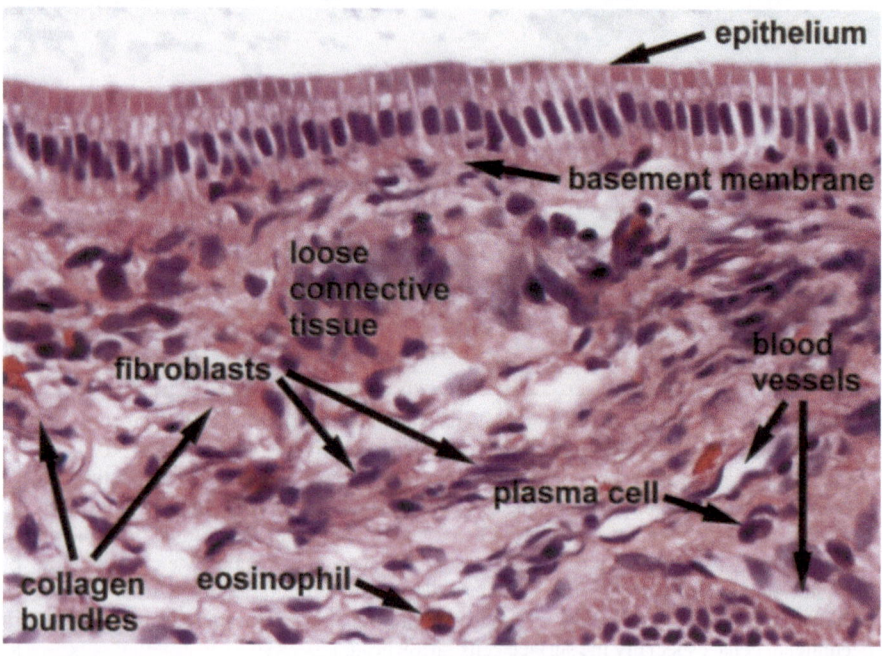

Figure 11. Loose (areolar) connective tissue underlies most epithelial layers an is often referrd to as the lamina propria. Loose connective tissue (CT) underlies the surface epithelium of the lumen of the gall bladder, and is separated from the epithelium by a basement membrane. Loose CT is basically composed of cells and extracellular matrix, which is composed of ground substance and collagen fibers (pink). The collagen is produced by the fibroblasts. There are many small blood vessels, and a wide range of permanent and immigrant cells, such as lymphocytes, macrophages, plasma cells, and leukocytes such as neutrophils and eosinophils.

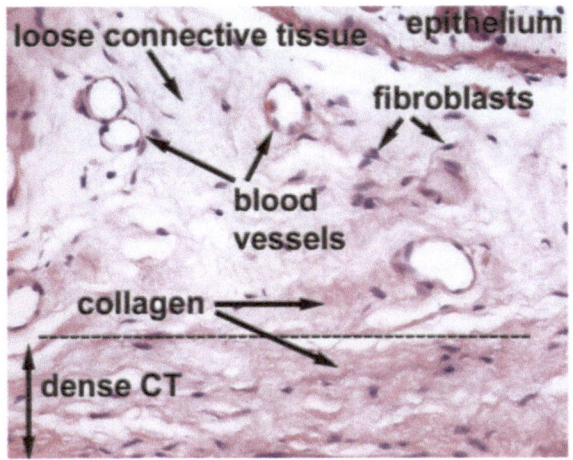

Figure 12. Loose connective tissue. The loose connective tissue (CT) underlies the surface epithelium of the lumen of the stomach, and is composed of a loose matrix of collagen fibers, extracellular matrix and fibroblasts There are many small blood vessels, and deep to the loose CT is dense irregular CT.

II.I.II   Dense connective tissue
A. Dense irregular connective tissue

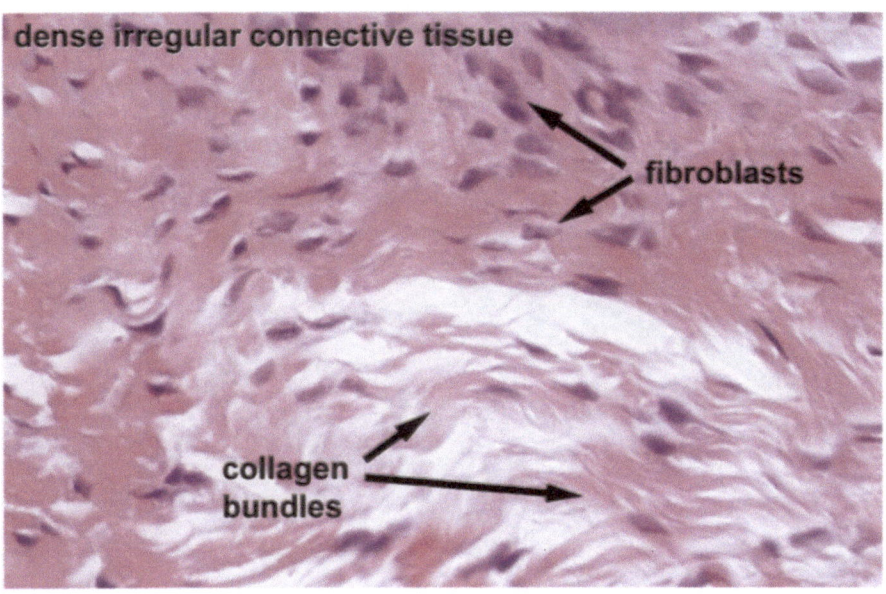

Figure 13. Dense irregular connective tissue. Dense irregular connective tissue (CT) surrounds hyaline cartilage, and is composed of a dense matrix of large collagen bundles, extracellular matrix and fibroblasts There are very few blood vessels In this specimen, some of the cells may be chondroblasts. Another example of dense irregular CT is shown at the bottom of figure 11.

B.   Dense regular connective tissue

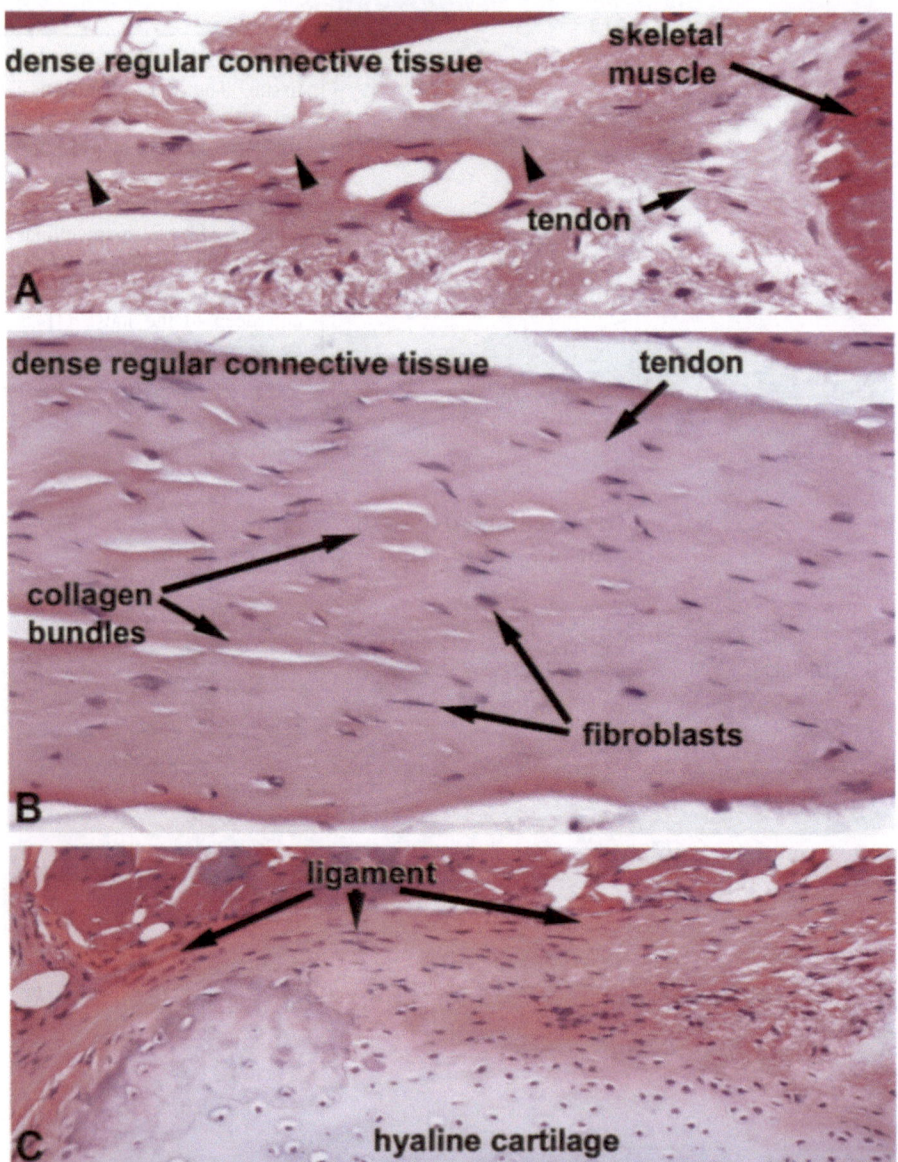

Figure 14. Dense regular connective tissue. **A.** Dense regular connective tissue (CT) is predominant in tendons. Note the attachment of the tendon to the skeletal muscle. **B.** High magnification of a tendon showing collagen bundles oriented parallel to each other, with fibroblasts interspersed in parallel with the collagen bundles. **C.** Dense regular CT is predominant in ligaments. Ligaments resemble tendons, but the fibroblasts tend to have a more regular orientation with the collagen bundles.

II.II   Specialized connective tissues
II.II.I   Cartilage

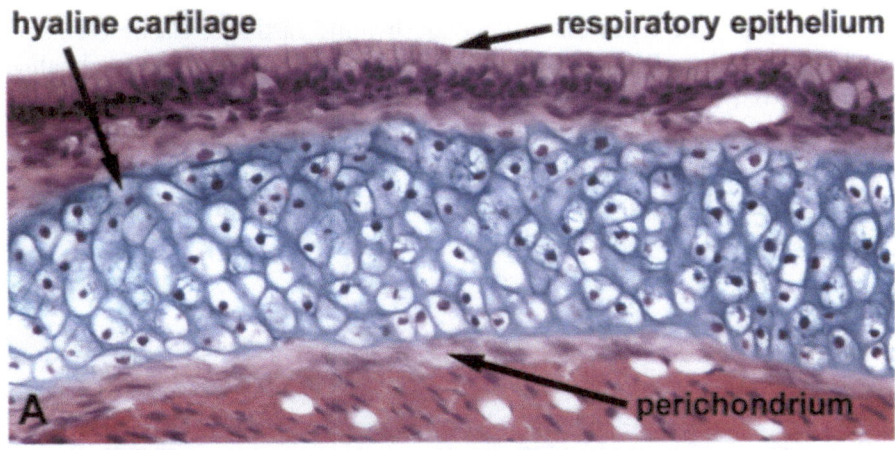

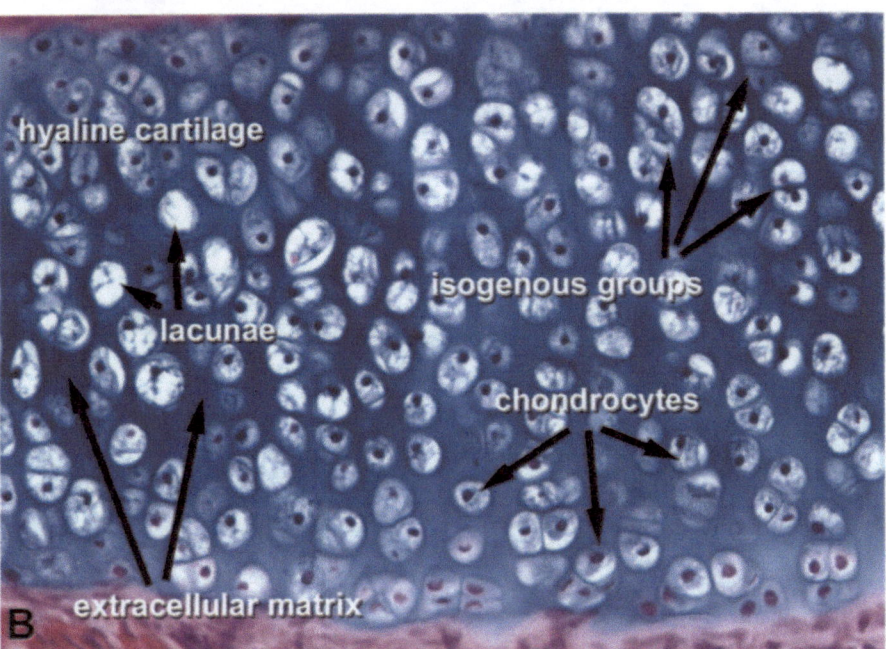

Figure 15. Hyaline cartilage. **A.** Low magnification image of the hyaline cartilage that provides the structural support for the trachea. Note that the cartilage is completely surrounded by a dense connective tissue perichondrium, which is one source of new chondrocytes. **B.** High magnification the trachea showing the lacunar spaces occupied by chondrocytes. Chondrocytes are not well-preserved during fixation, but the nuclei can be discerned. Isogenous groups are clusters of daughter chondrocytes that have divided, and will produce additional basophilic matrix resulting in interstitial growth.

Cartilage is a specialized connective tissue that provides structural support. The three types of cartilage are hyaline cartilage, elastic cartilage, and fibrocartilage. Elastic cartilage is similar in structure to hyaline cartilage, but has elastic fibers in the matrix in addition to collagen fibers. Fibrocartilage is often attached to bone as a transition tissue between dense connective tissue and bone or hyaline cartilage. There are many large collagen bundles in the extracellular matrix of fibrocartilage, and fibrocartilage is not surrounded by a perichondrium.

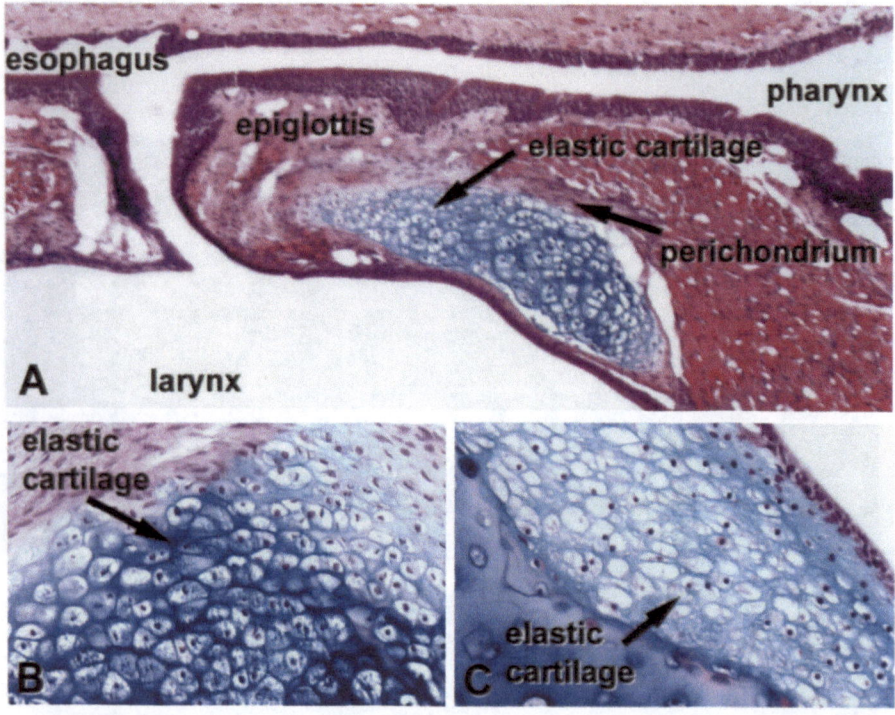

Figure 16. Elastic cartilage. **A.** Low magnification image of the elastic cartilage that provides the structural support for the epiglottis. Note that the cartilage is completely surrounded by a perichondrium, which is one source of new chondrocytes. **B.** High magnification of the epiglottis showing the lacunar spaces occupied by chondrocytes. The presence of the thin branching elastic fibers cannot be easily observed in H & E preparations. Isogenous groups are present, as in hyaline cartilage (figure 14). **C.** Elastic cartilage provides flexible structural support for the vocal chords.

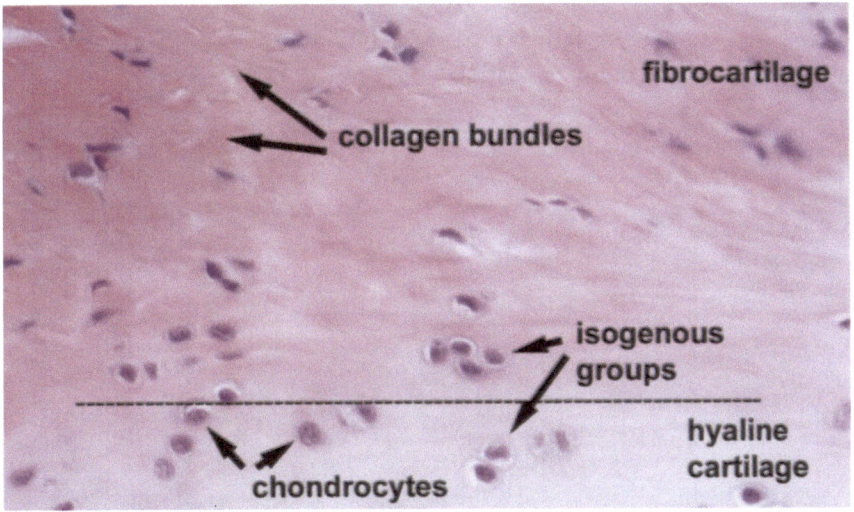

Figure 17. High magnification image of the fibrocartilage that provides a transition zone between dense regular connective tissue and hyaline cartilage. Fibrocartilage does not have a perichondrium, so all of its growth is interstitial (within the cartilage). Lacunar spaces are occupied by chondrocytes, and isogenous groups are present. There are large bundles of eosinophilic collagen in the extracellular matrix.

II.II.II   Bone

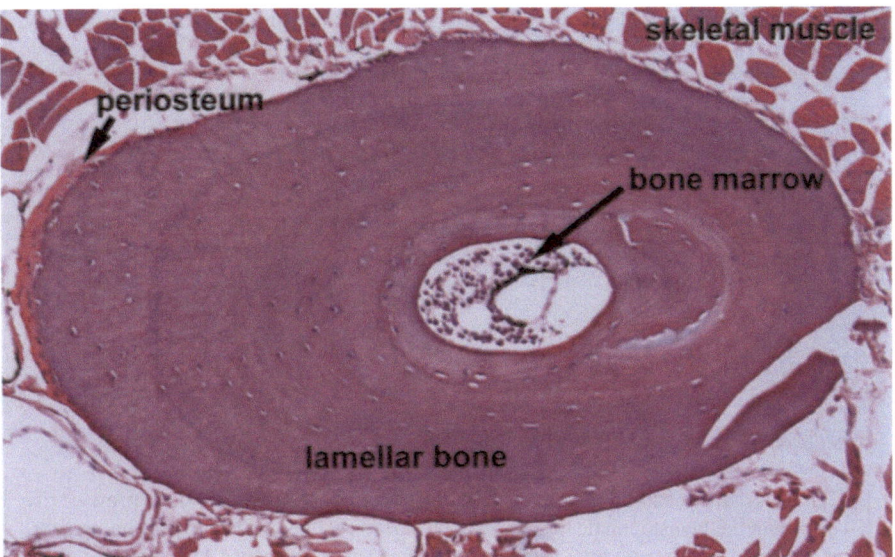

Figure 18. Low magnification image of the mature lamellar bone of a limb. In the center of the bone is the marrow containing hematopoietic cells. Osteocytes are located in lacunae in a circular pattern around the circumference.

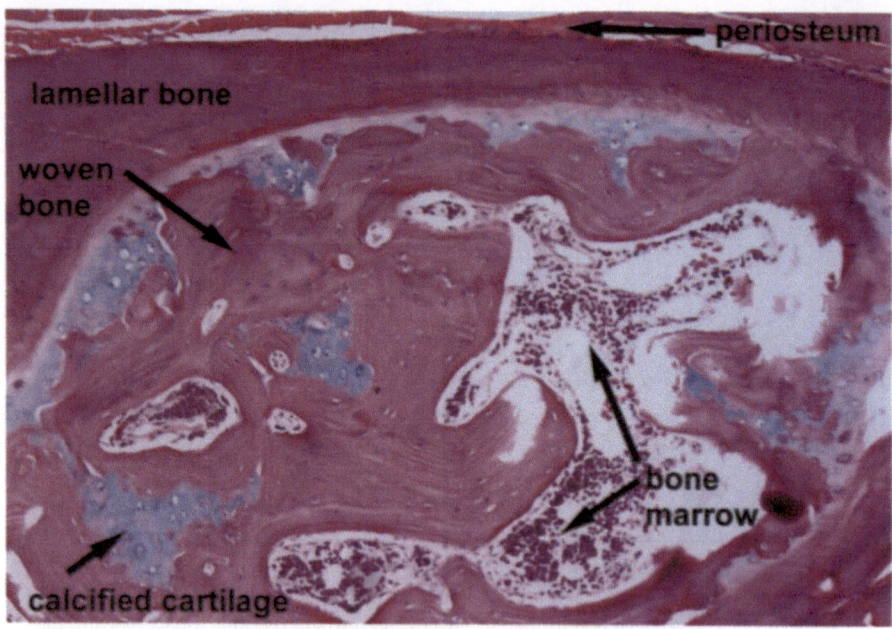

Figure 19. Lamellar and woven bone. A low magnification image of a bone in cross-section showing the mature lamellar bone surrounding the outer circumference and covered with a periosteum. The woven or immature bone lies deeper in the bone organ, and the lacunae are oriented randomly rather that in a circumferential pattern as in the lamellar bone. In the center of the bone lies the marrow cavity containing the hematopoietic cells. Calcified cartilage is present, which is converted to woven bone by osteoblasts.

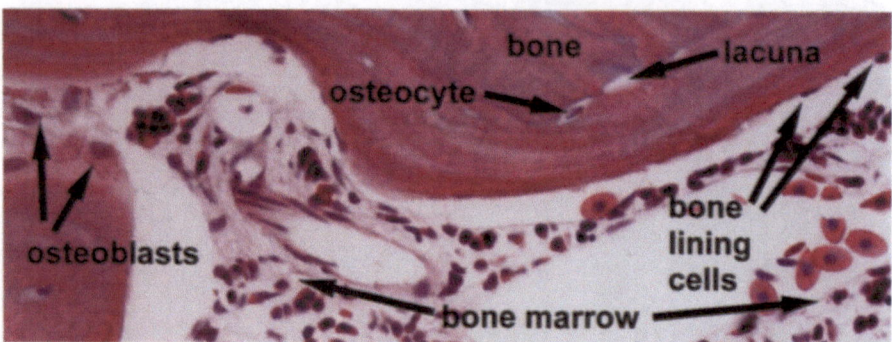

Figure 20. Bone and marrow cavity. In the center of the bone lies the marrow cavity containing the hematopoietic cells. The bone surface is lined with flat bone lining cells, which can differentiate into osteoblasts as needed. Osteoblasts also line the bone surface, and produce and calcify the bone matrix. Osteocytes are located in lacunae, and communicate with other osteocytes by small cell processes located in small channels in the bone called canaliculi.

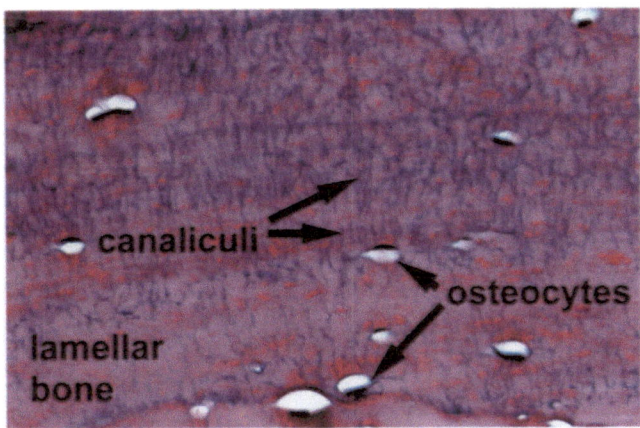

Figure 21. Canaliculi in lamellar bone. The thin blue lines are small channels called canaliculi that radiate from osteocytes and contain cellular processes of adjacent cells which contact each other via gap junctions. These communicating junctions are critical for the passage of oxygen and nutrients to osteocytes.

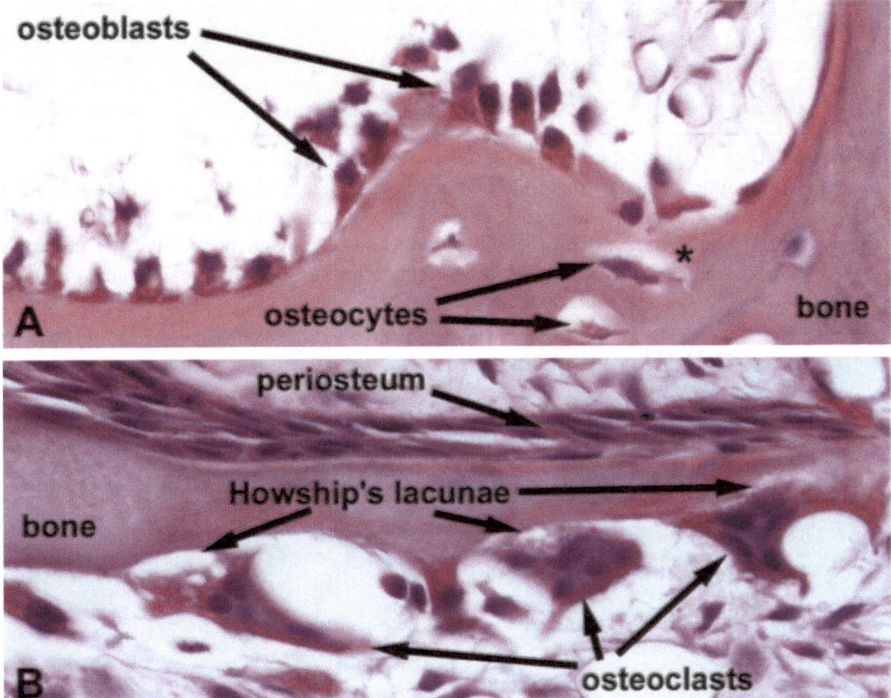

Figure 22. Osteoblasts and osteoclasts. **A.** Osteoblasts line the bone surface, and produce and calcify the bone matrix (osteoid). They are basophilic cuboidal-like cells. After secreting osteoid, they eventually become entrapped in the bone matrix that they produced and become osteocytes (*). **B.** Osteoclasts resorb bone by expelling lysosomal enzymes and carbonic acid onto the bone surface. The osteoclasts are closely adhered to the bone, and the area of bone resorption is called a Howship's lacuna. A dense connective tissue called the periosteum surrounds the outer surface of bone.

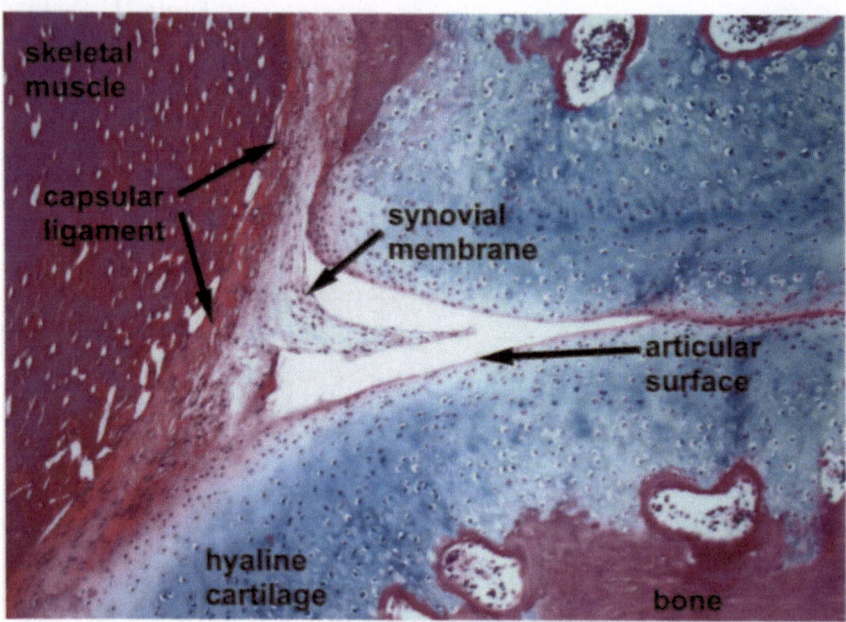

Figure 23. Low magnification image of a synovial joint. The ends of the long bones are covered with hyaline cartilage where the two bones articulate with each other to form a joint. There is no perichondrium on the articulating surface of the hyaline cartilage. The synovial joint is enclosed within a synovial capsule, with a capsular ligament forming the outermost layer. A delicate connective tissue synovial membrane lines the synovial cavity containing the synovial fluid made by cells of the synovial membrane.

II.II.III Adipose

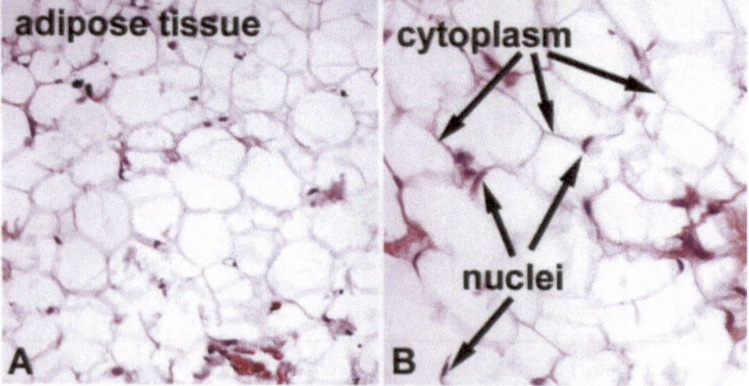

Figure 24. Adipose tissue (white fat). **A.** Low magnification of adipose tissue. The white space taking up most of the interior of the cells are the location of the fat bodies, which were extracted during tissue processing. **B.** High magnification of adipose tissue. The adipose cells have a thin rim of eosinophilic cytoplasm, and a dense eccentric nucleus.

II.II.IV   Blood

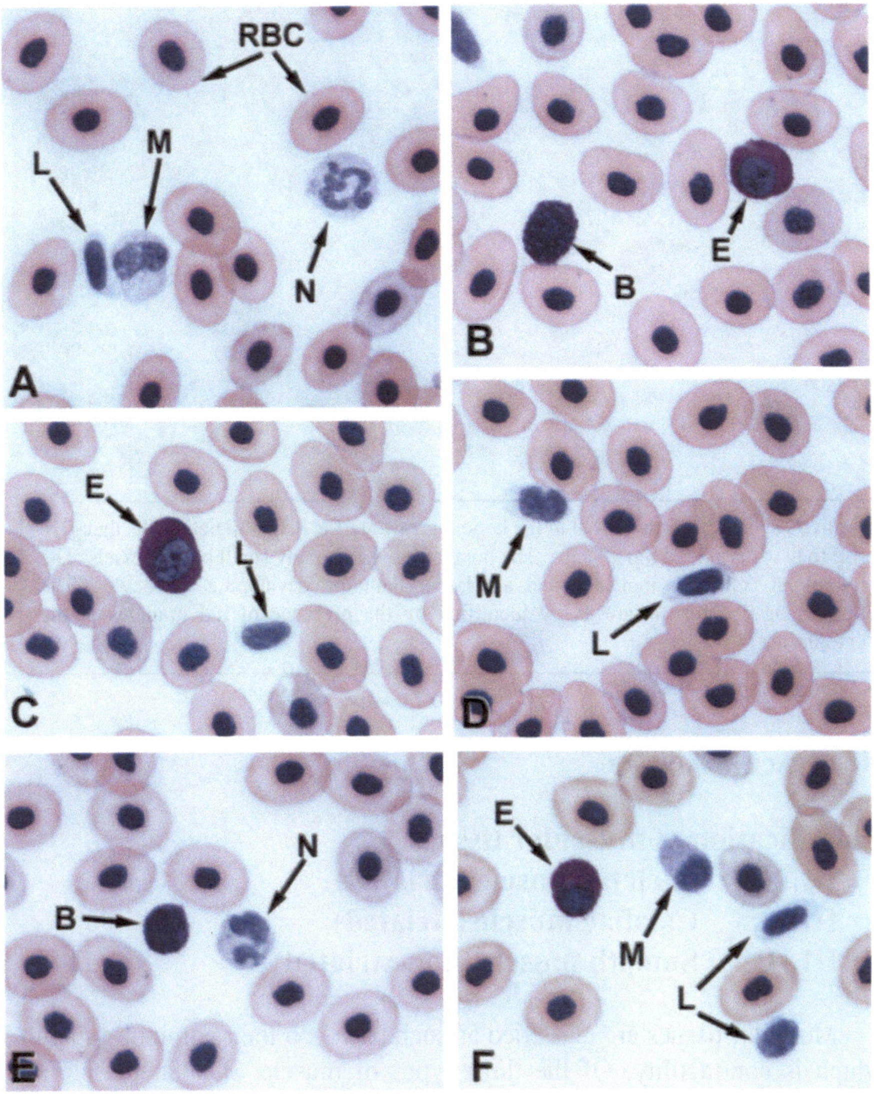

Figure 25. Peripheral blood smear stained with Wright's stain. **A.** Red blood cells (RBCs; erythrocytes) are nucleated, with eosinophilic cytoplasm, and are larger than the leukocytes. Neutrophils (N) have a relatively pale cytoplasm, with a highly lobulated nucleus. Monocytes (M) have a large indented nucleus with pale blue cytoplasm, and lymphocytes (L) have small oval nuclei with blue cytoplasm. **B.** Basophils (B) have dark blue granules, and eosinophils (E) have deep red granules in the cytoplasm. **C-F.** Additional examples of leukocytes and RBCs. B, basophil; E, eosinophil; L, lymphocyte; M, monocyte; N, neutrophil.

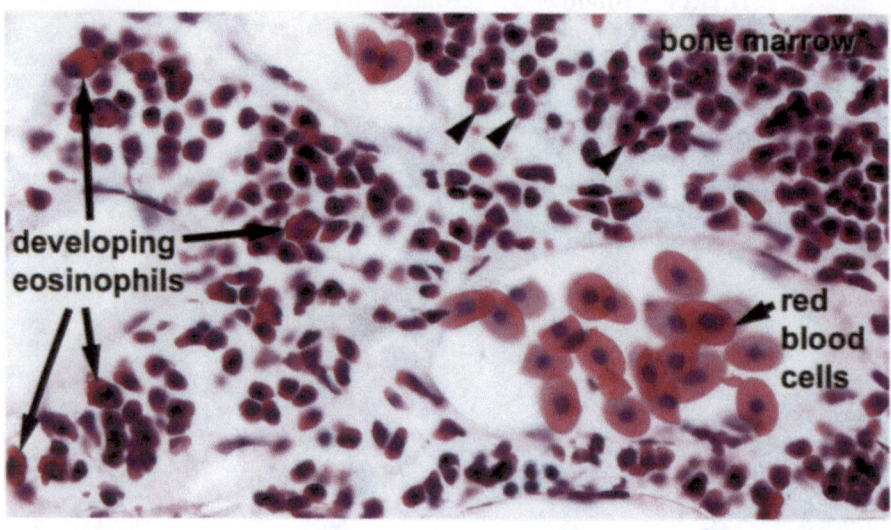

Figure 26. Bone marrow. Within the bone marrow are the hematopoietic cells that give rise to new peripheral blood cells. The bone marrow has many small blood vessels. The red blood cells are nucleated and are larger than the developing leukocytes. The developing eosinophils are easily identified by the presence of red granules in the cytoplasm.

III Muscular Tissue

Classification of muscular tissues
    III.I   **Skeletal muscle (striated)**
    III.II   **Cardiac muscle (striated)**
    III.III   **Smooth muscle (non-striated)**

Muscular tissues are classified as such based on their common function, which is contractility. Of the three types of muscle, skeletal muscle and cardiac muscle can be classified as striated muscle because of the presence of striations perpendicular to the long axis of the cells. The striations are the result of the highly organized myofilaments that are responsible for muscle contraction.

Skeletal muscle cells are actually syncytiums of many cells, so they are referred to as muscle fibers, and the plasma membrane and cytoplasm are referred to as the sarcolemma and sarcoplasm, respectively. Cardiac muscle cells are highly branches, and are connected to each other by junctional complexes. Smooth muscle cells are relatively smaller, and their contractions are generally under autonomic control.

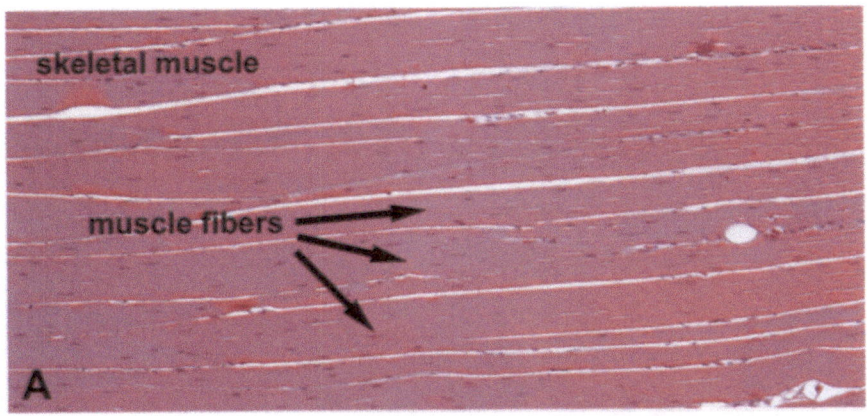

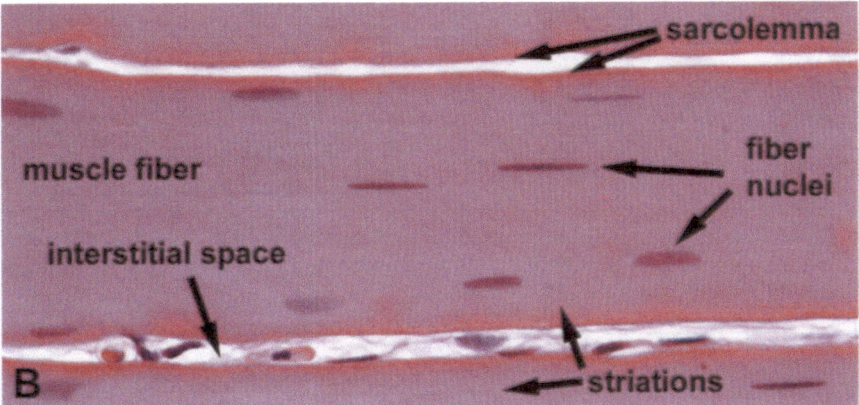

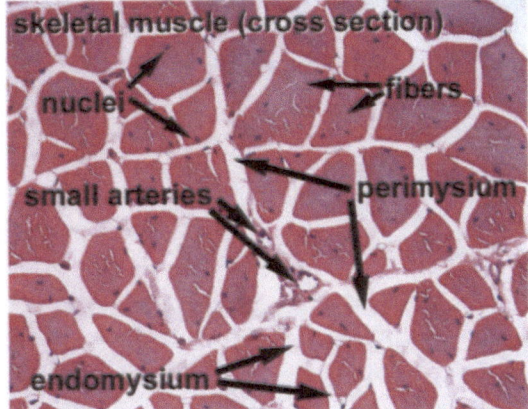

Figure 27. (ABOVE) Longitudinal section of skeletal muscle. **A.** Low magnification of skeletal muscle indicating the individual fibers. **B.** High magnification of muscle fibers. Interstitial space surrounding the fibers contain blood vessels and loose connective tissue. Note the alternating dark and light bands perpendicular to the long axis of the fibers.

Figure 28. Cross section of skeletal muscle. Low magnification of the organization of the muscle fibers. Fiber bundles are separated from each other by a connective tissue perimysium, and the individual fibers are separated by an endomysium. Fiber nuclei are located randomly throughout the sarcoplasm.

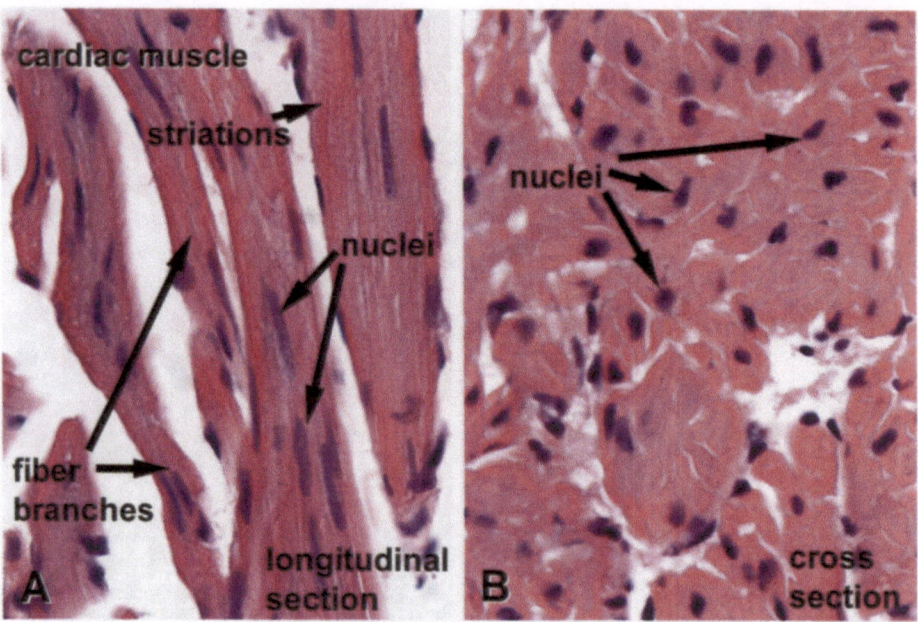

Figure 29. Cardiac muscle. **A.** Longitudinal section of cardiac muscle. Cells are branched, and have striations in the cytoplasm. **B.** Cross section of cardiac muscle. The nuclei are located centrally within the cells.

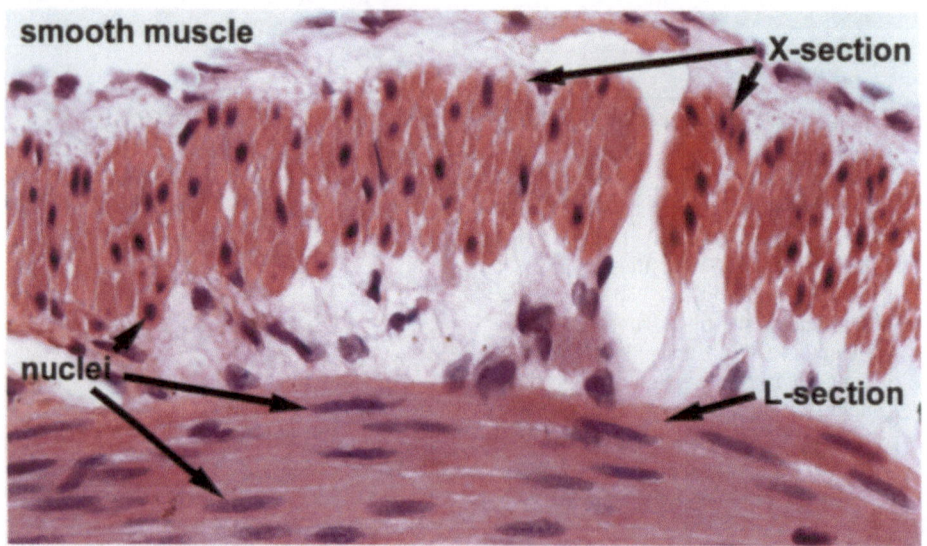

Figure 30. Smooth muscle. Longitudinal (L) and cross (X) section of smooth muscle in the muscularis externa of the colon. Note the fusiform (tapered) shape of the cells.

IV Nervous Tissue

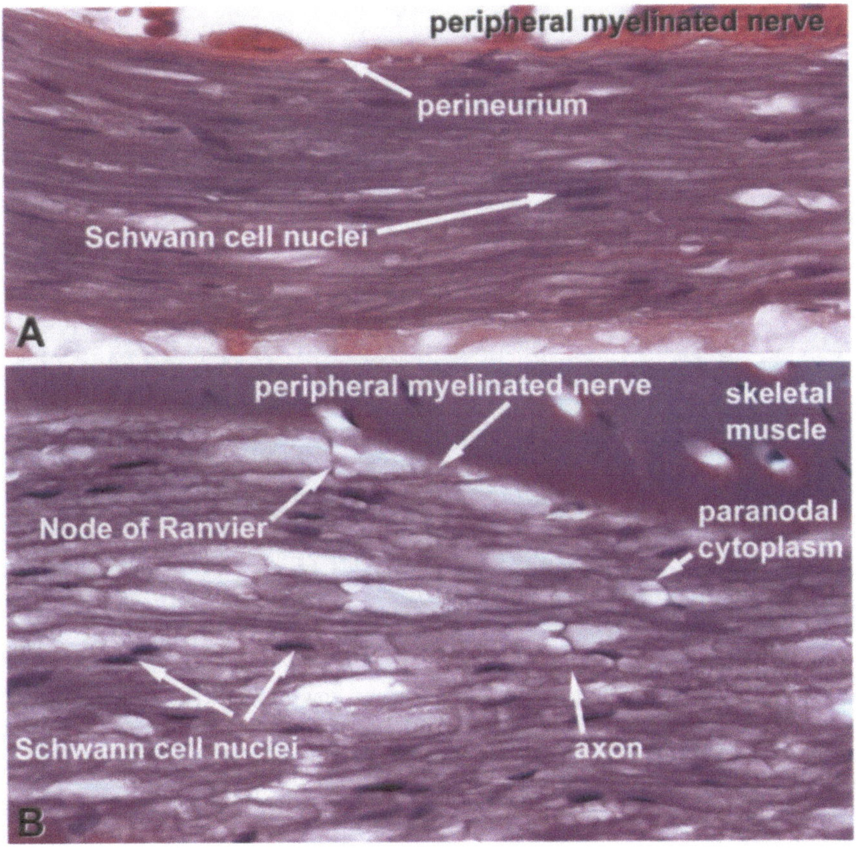

peripheral myelinated nerve

perineurium

Schwann cell nuclei

A

peripheral myelinated nerve

skeletal muscle

Node of Ranvier

paranodal cytoplasm

Schwann cell nuclei

axon

B

Figure 31. Myelinated peripheral nerve. **A.** Low magnification of peripheral nerve showing the cellular perineurium that surrounds the nerve bundle. Note the elongate nuclei of the myelin-producing Schwann cells. **B.** High magnification of peripheral nerve. Axons are ensheathed in layers of myelin The Nodes of Ranvier are located between the paranodal cytoplasm.

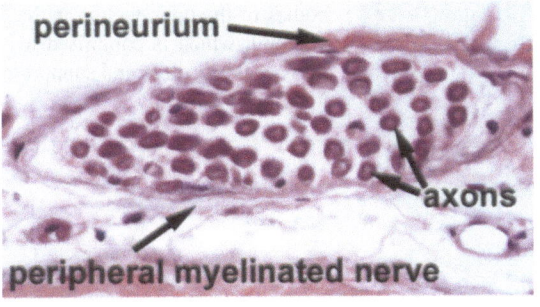

perineurium

axons

peripheral myelinated nerve

Figure 32. Myelinated peripheral nerve in cross section. Small axons are surrounded by the layers of myelin. The nerve bundle is surrounded by the perineurium.

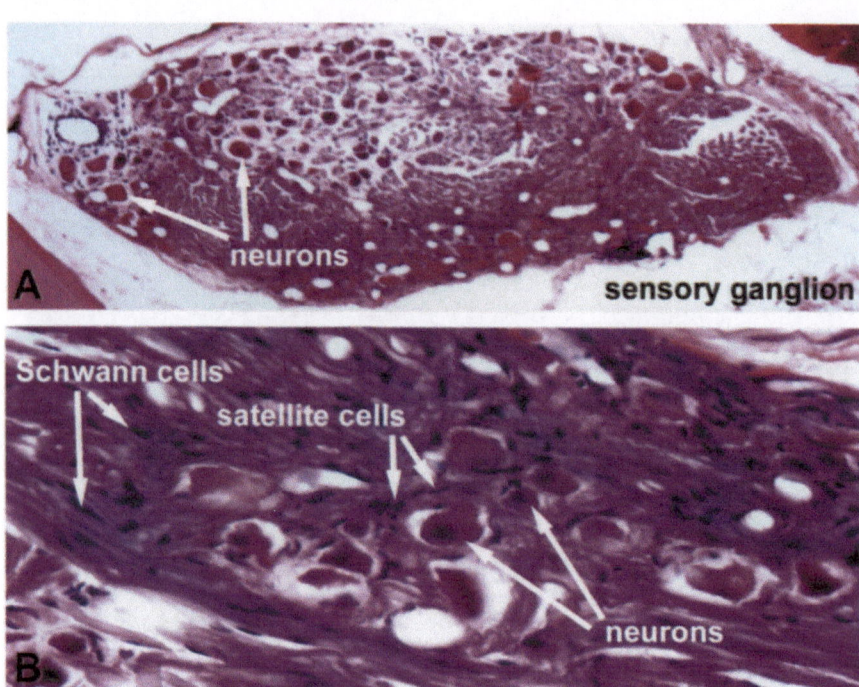

Figure 33. Sensory (trigeminal) ganglion. **A.** Low magnification of sensory ganglion. The ganglion has many large pseudounipolar neurons. **B.** High magnification of sensory ganglion. The sensory neurons are surrounded by satellite cells. Myelinated nerves pass through the ganglion.

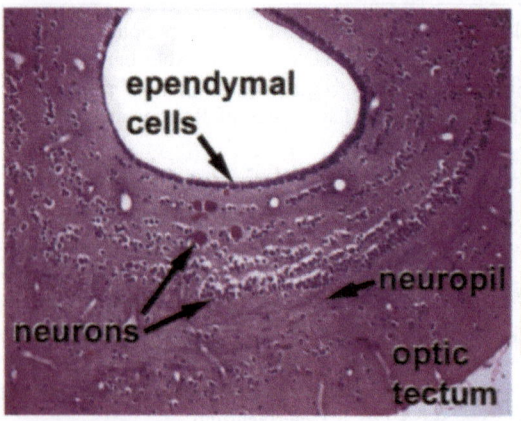

Figure 34. Neurons in the central nervous system. The brain tissue that is in contact with the ventricles is covered with a single layer of ependymal cells. Surrounding the neural cell bodies of the optic tectum is the neuropil, which is comprised of nerve cell processes and supporting glial cells.

Chapter 2

Cardiovascular System

The cardiovascular system consists of the heart and blood vessels and distributes oxygen, nutrients, and other molecules to the tissues and collects waste products. Thin-walled lymphatic vessels transport lymphatic fluid from the tissues into the circulation.

I   BLOOD VESSELS

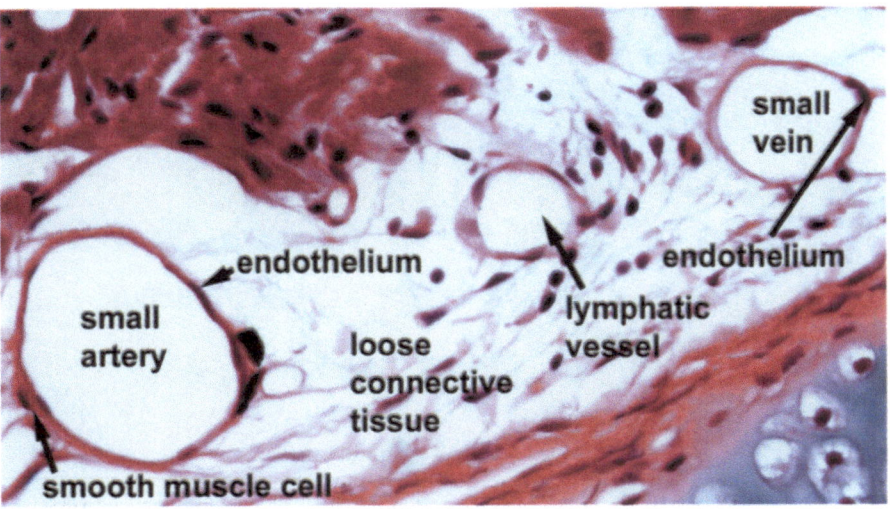

Figure 1. Small artery, vein, and lymphatic vessel. A small artery and small vein are seen in the loose connective tissue underlying the cartilage of the trachea. The small veins consist almost entirely of a single layer of endothelium (simple squamous epithelium), whereas the arteries have at least one layer of smooth muscle overlying the endothelium. The lymphatic vessel resembles the vein, but can be distinguished based on the appearance of the endothelium.

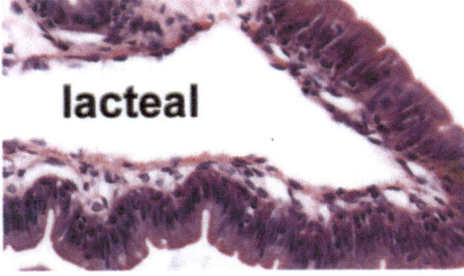

Figure 2. Lymphatic vessels. Lymphatic vessels begin as blind-ended sacs (lacteals) that collect lymphatic fluid from the interstitial space of tissues, such as in the lamina propria of the duodenum shown here. These vessels are lined by an endothelium, and are often difficult to discern from veins.

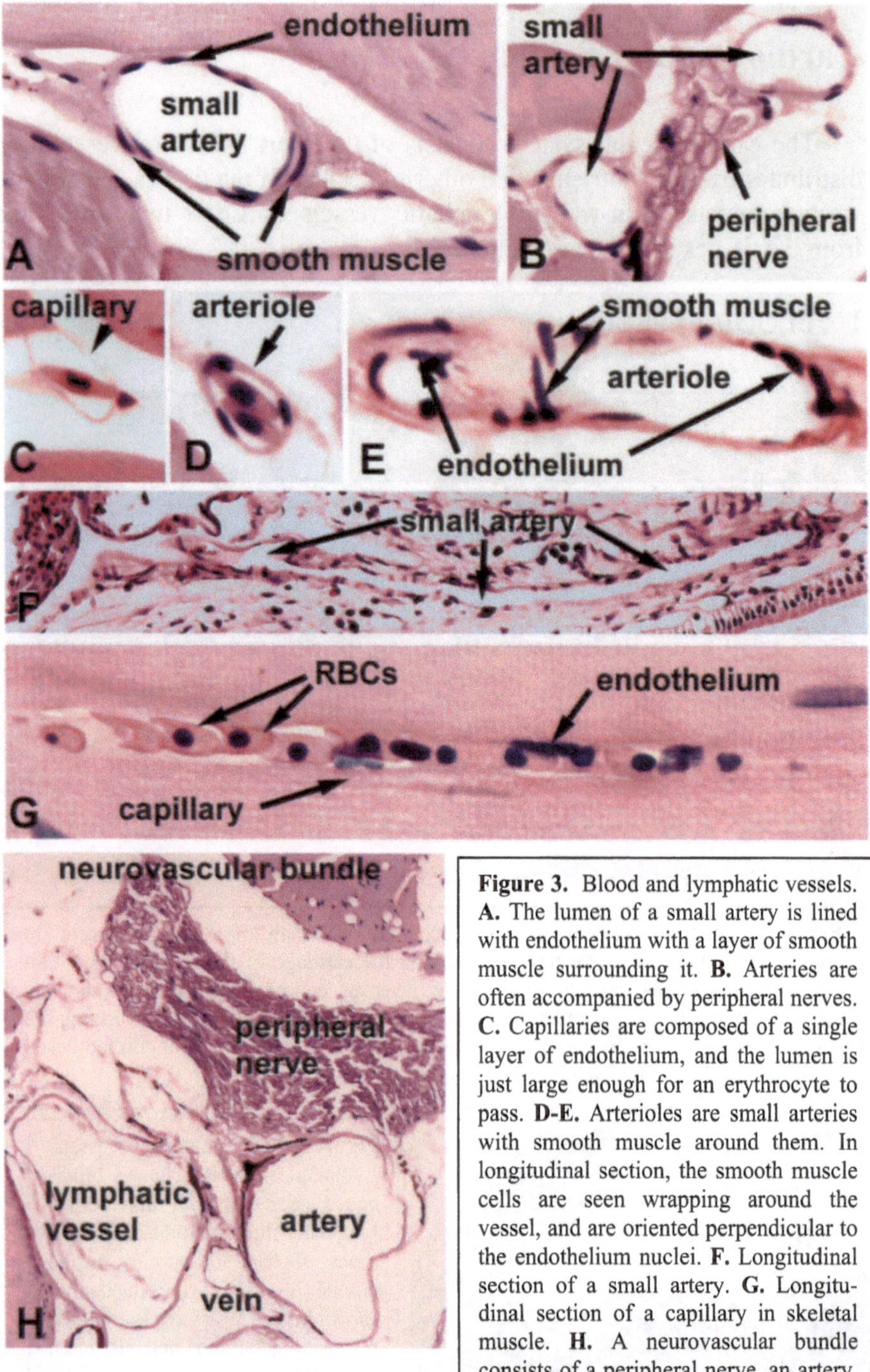

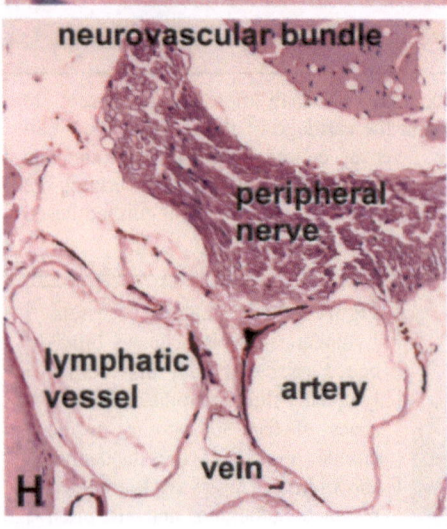

Figure 3. Blood and lymphatic vessels. **A.** The lumen of a small artery is lined with endothelium with a layer of smooth muscle surrounding it. **B.** Arteries are often accompanied by peripheral nerves. **C.** Capillaries are composed of a single layer of endothelium, and the lumen is just large enough for an erythrocyte to pass. **D-E.** Arterioles are small arteries with smooth muscle around them. In longitudinal section, the smooth muscle cells are seen wrapping around the vessel, and are oriented perpendicular to the endothelium nuclei. **F.** Longitudinal section of a small artery. **G.** Longitudinal section of a capillary in skeletal muscle. **H.** A neurovascular bundle consists of a peripheral nerve, an artery, a vein, and a lymphatic vessel.

II HEART

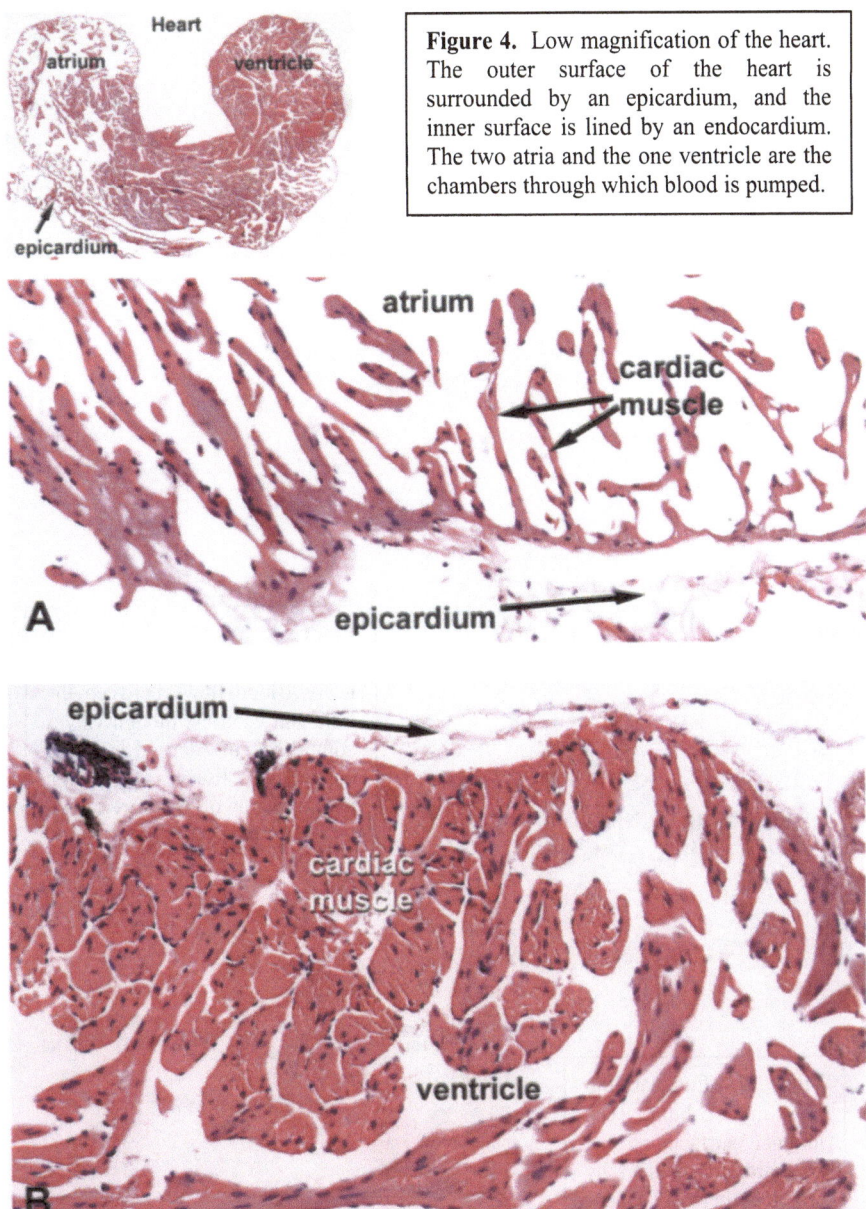

Figure 4. Low magnification of the heart. The outer surface of the heart is surrounded by an epicardium, and the inner surface is lined by an endocardium. The two atria and the one ventricle are the chambers through which blood is pumped.

Figure 5. The atrium and ventricle. **A.** The atria are surrounded by an epicardium that consists of a layer of endothelium and some loose connective tissue containing blood vessels and nerves. The myometrium, composed of cardiac muscle is organized into thin strands. **B.** The myometrium of the ventricle is quite large and is also surrounded by an epicardium.

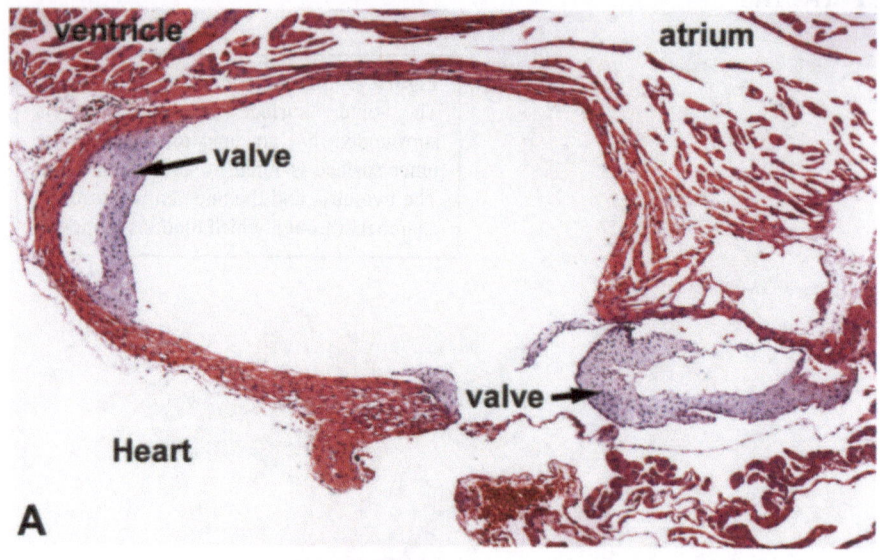

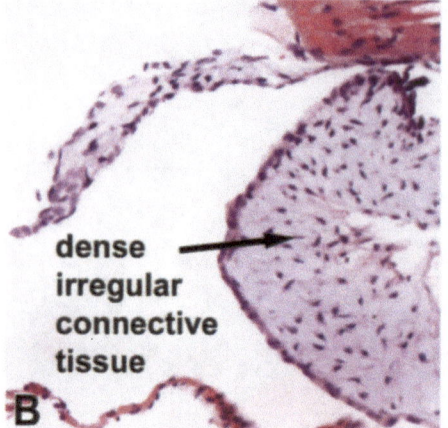

Figure 6. Heart valves. **A.** Low magnification of the valve of the heart. Blood flows between the atria and ventricle though valves composed of dense irregular connective tissue. **B.** High magnification of the heart valve.

Figure 7. Endocardium and myocardium. The luminal surface of the myocardium is lined by an endothelium (endocardium; simple squamous epithelium).

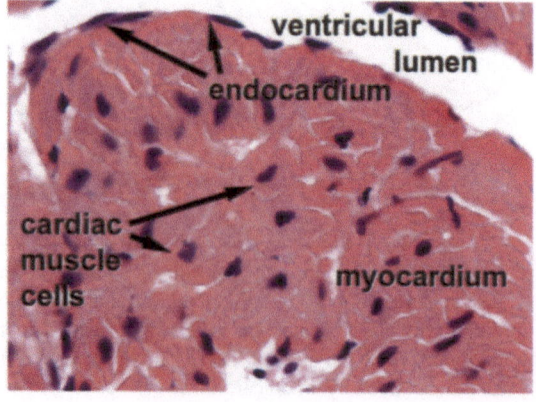

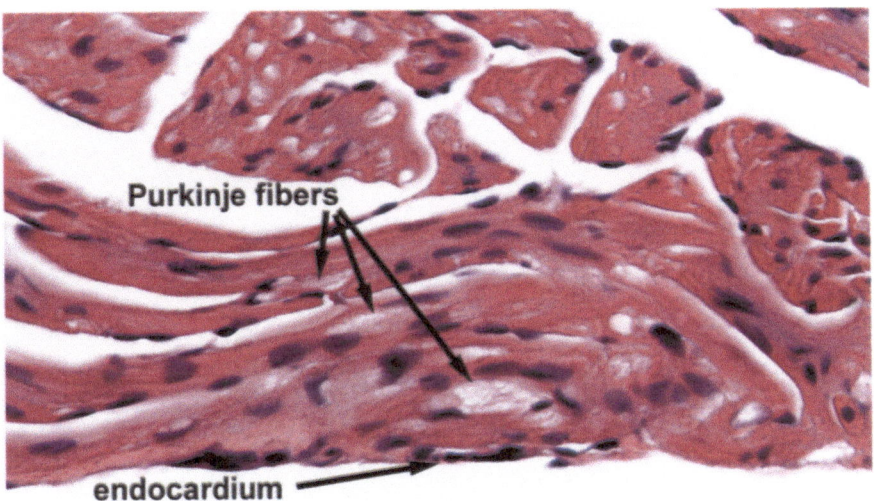

Figure 8. Purkinje fibers. Located in the subendocardial region of the ventricle are Purkinje fibers. They are modified cardiac muscle cells specialized for the conduction of electric impulses from the atrioventricular node. The abundance of glycogen in the cytoplasm of Purkinje fibers accounts for their lighter-staining appearance, since the glycogen is extracted during tissue preparation.

Chapter 3

Lymphatic organs

The lymphatic system consists of groups of cells and organs organized to defend the body against foreign antigens, such as bacteria, viruses, and tumor cells. The cells are mostly lymphocytes and macrophages, as well as the connective tissue cells that provide structural support for the organs.

I DIFFUSE LYMPHOID TISSUE

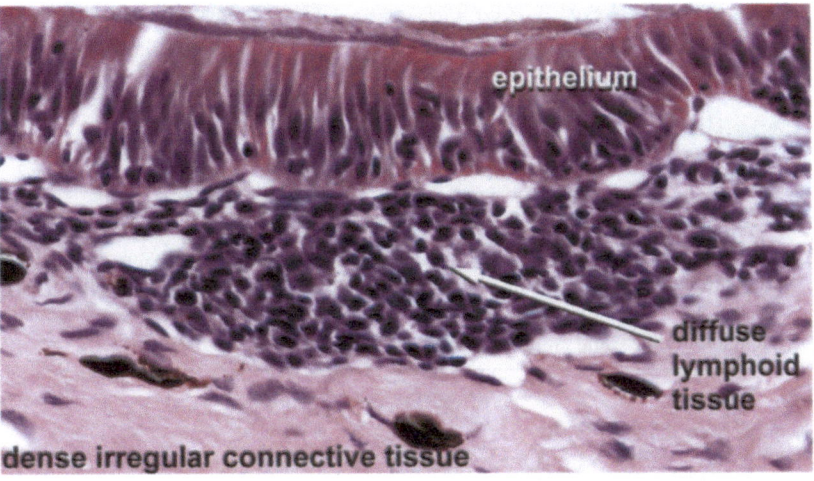

Figure 1. Diffuse lymphoid tissue. Diffuse lymphoid tissues are localized collections of lymphocytes, and are common in the lamina propria of the alimentary and respiratory tracts. They are supported by a stroma of reticular fibers and reticular cells.

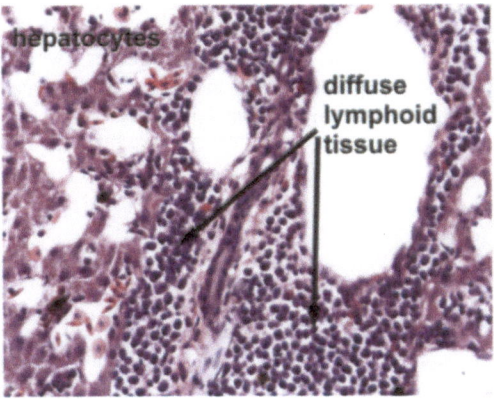

Figure 2. Diffuse lymphoid tissue. In the liver, diffuse lymphoid tissues are located in the loose connective tissue associated with the portal triads.

II LYMPH NODE

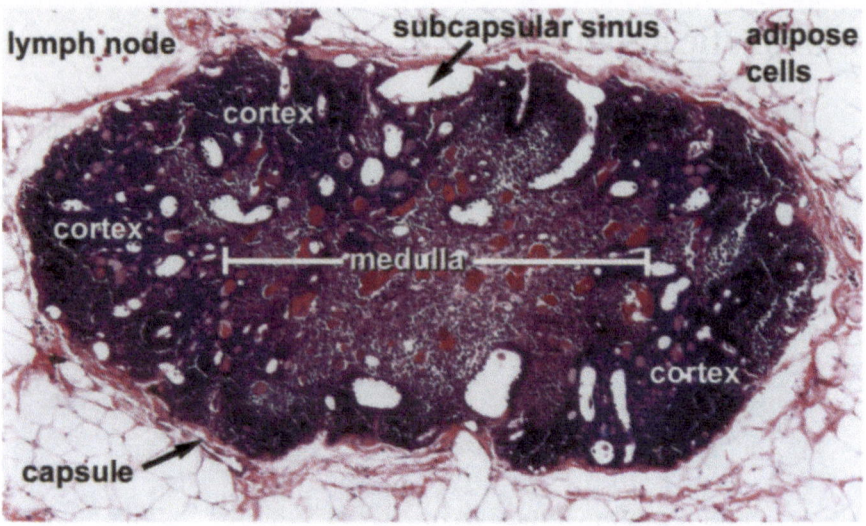

Figure 3. Lymph nodes of *Xenopus*. Lymph nodes are small organs occurring in series along lymphatic vessels. They are surrounded by a dense connective tissue capsule. The lymphoid tissue is in two regions: the outer cortex which contains lymphatic nodules, and an inner medulla which contains medullary cords.

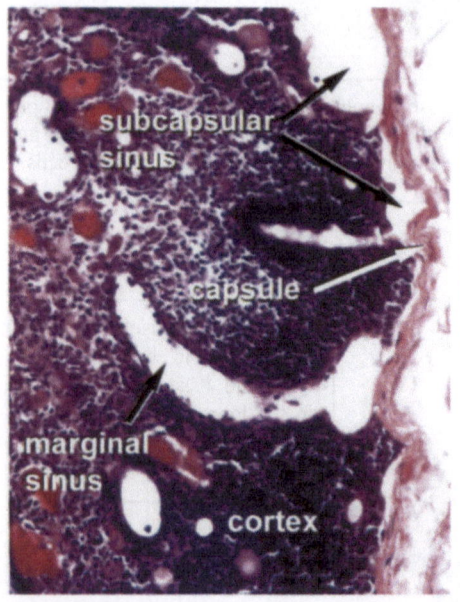

Figure 4. Sinuses of the lymph node cortex. Underlying the capsule is a sinus that receives lymphatic fluid and cells from the afferent lymphatic vessels. The subcapsular sinuses drain into the marginal sinuses which penetrate into the cortex to give rise to the medullary sinusoids.

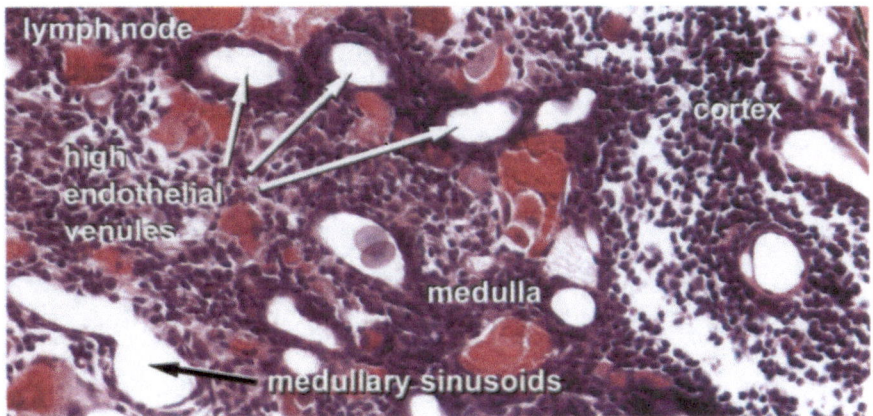

Figure 5. High endothelial venules. In the area of the cortex adjacent to the medulla are located high endothelial venules (HEVs). Lymphocyte movement from the circulatory system into the lymph nodes occurs at the HEVs, which are lined by unusually tall endothelial cells.

III SPLEEN

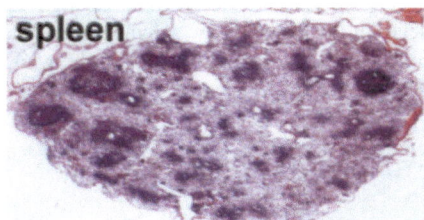

Figure 6. Low magnification of the spleen. Note the distribution of the dense-staining lymphatic nodules (white pulp) containing many lymphocytes. Surrounding the white pulp is the lighter-staining red pulp that is composed of splenic cords and sinuses.

Figure 7. Red and white pulp of the spleen. Located near the center of the white pulp is a central artery which conducts blood towards the splenic sinuses, which are located in the red pulp.

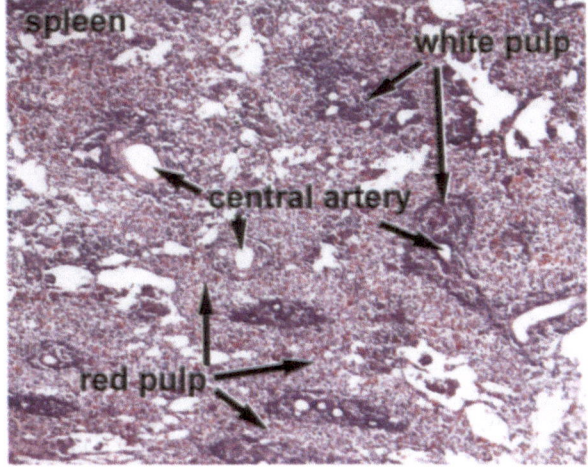

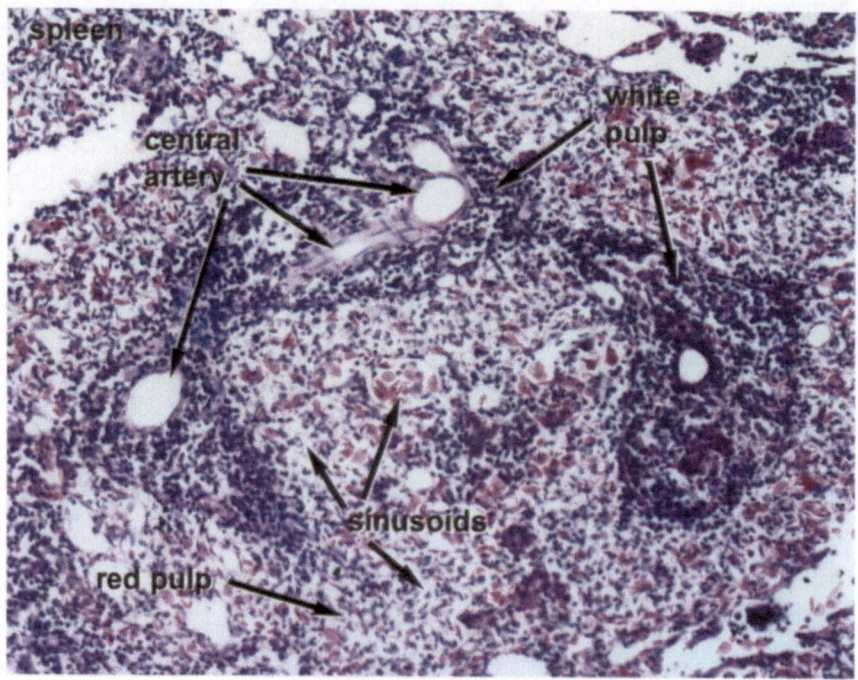

Figure 8. Red and white pulp of the spleen. In the red pulp, the splenic sinuses are filled with erythrocytes and other cells of the general circulation. Within the white pulp are lymphocytes and macrophages.

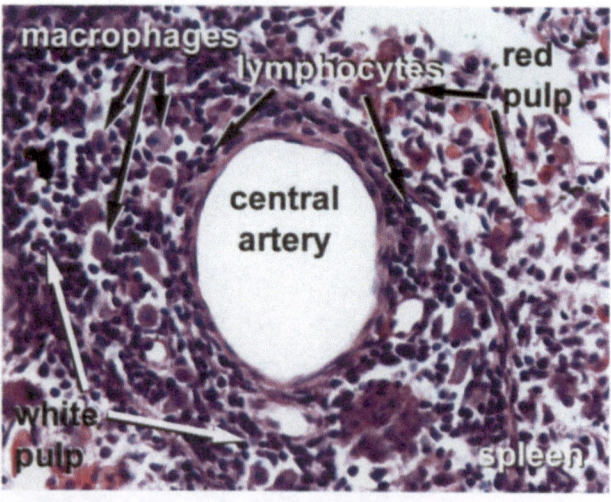

Figure 9. High magnification of the white pulp of the spleen. The lymphocytes surround the central artery. Macrophages are also prominent in the white pulp and can be identified by the presence of eosinophilic cytoplasm.

Chapter 4

Gastrointestinal System

The major function of the gastrointestinal system is to digest and absorb nutrients from ingested material. The ingested material travels down a series of tubes (esophagus, stomach, intestines), and accessory organs (liver, gall bladder, spleen) secrete compounds into the digestive tubes to aid in digestion. The oral cavity and esophagus are covered in chapter 10.

I STOMACH

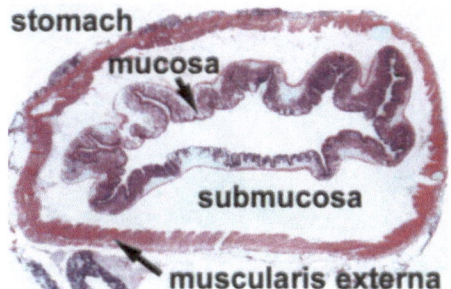

Figure 1. Low magnification image of the stomach. The lumen of the stomach is lined by a mucosa which is comprised of an epithelium, lamina propria, and muscularis mucosa. A muscularis externa is one of the most outer layers of the stomach.

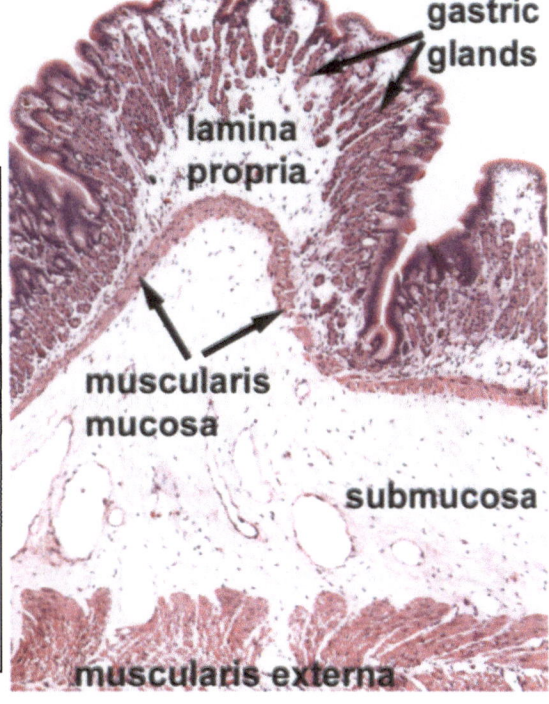

Figure 2. Major layers of the stomach. The mucosa is comprised of the surface epithelium, underlying lamina propria, and smooth muscle muscularis mucosa. Deep to the muscularis mucosa is the submucosa. Both the lamina propria and submucosa are comprised primarily of loose connective tissue. Deep to the submucosa is the muscularis externa, which is comprised of an inner circular and outer longitudinally-oriented layer of smooth muscle.

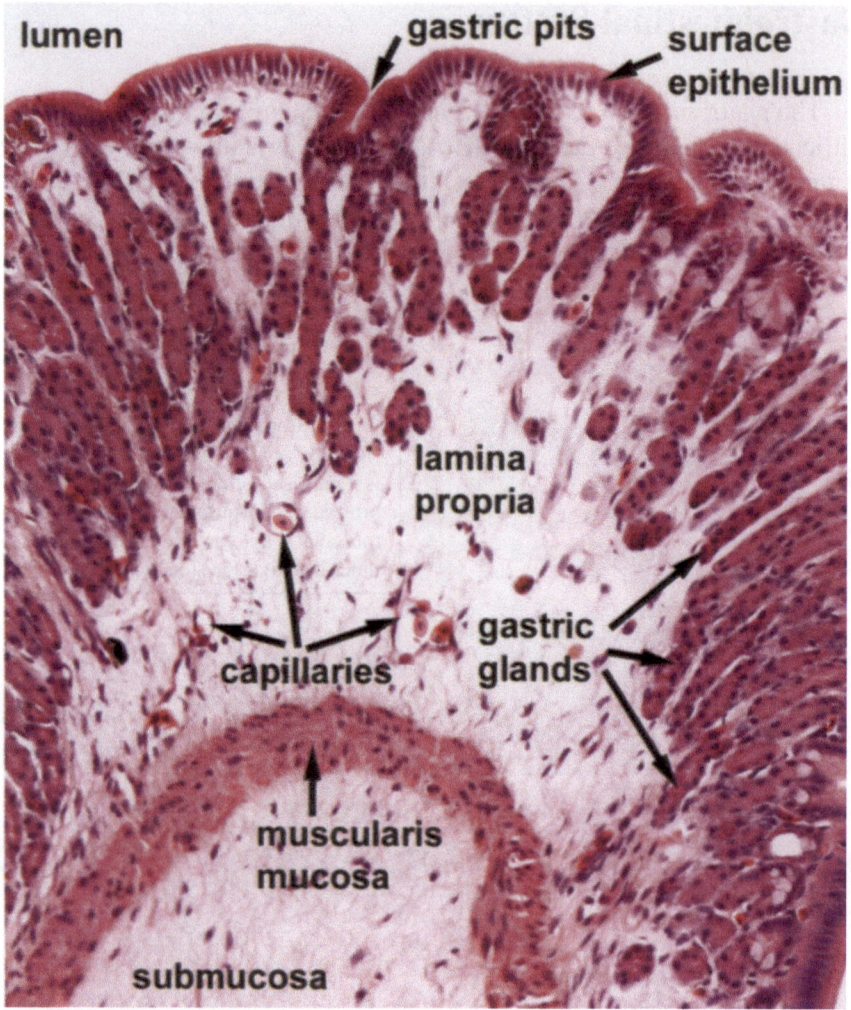

Figure 3. Mucous membrane of the stomach. The surface layer of the mucosa is composed of simple columnar epithelium. Invaginations of the surface epithelium form gastric pits, which also lines with surface epithelium. Branching off of the gastric pits are gastric glands that aid in digestion. The underlying loose connective tissue of the lamina propria contains many capillaries that transport absorbed nutrients to into the circulation. The deepest layer of the mucosa is the muscularis mucosa, which is a layer of smooth muscle cells.

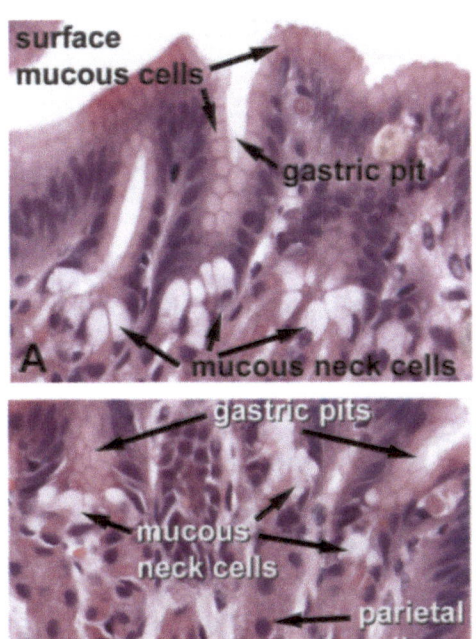

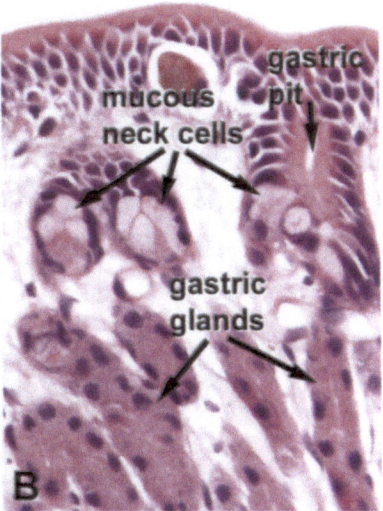

Figure 4. Epithelium and glands of the stomach. **A.** The surface epithelia are surface mucous cells and secrete mucus to protect the mucosa of the stomach. They invaginate into the lamina propria to form gastric pits. At the junction of the pits and glands are mucous neck cells (located in the neck of the glands). **B.** Branching from the gastric pits where the mucous neck cells are located, are the gastric glands, which extend deep into the lamina propria. **C.** The gastric glands in the fundic stomach are comprised primarily of parietal cells, which secrete hydrochloric acid to lower the pH of the lumen of the stomach.

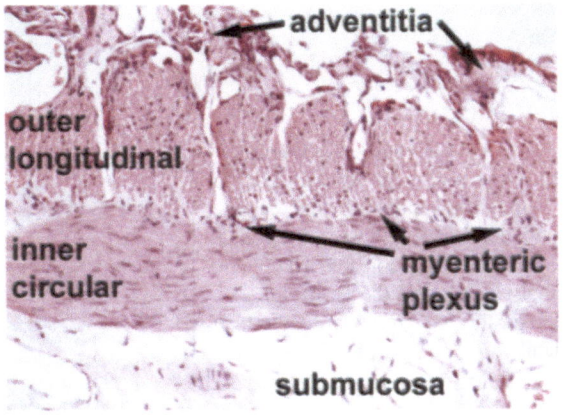

Figure 5. Muscularis externa of the stomach. Deep to the sub-mucosa are the inner circular and outer longitudindal layers of smooth muscle of the muscularis externa. Between the two layers of muscle is the myenteric plexus comprised of autonomic nerve ganglia, which innervates the muscularis externa to generate peristaltic movements.

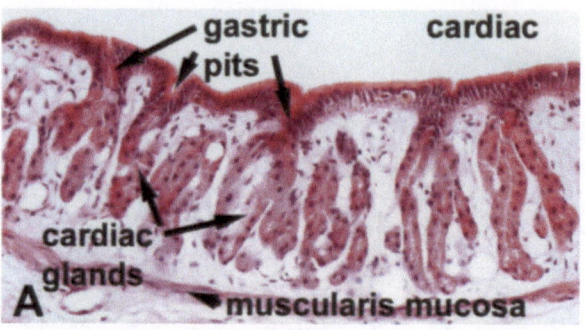

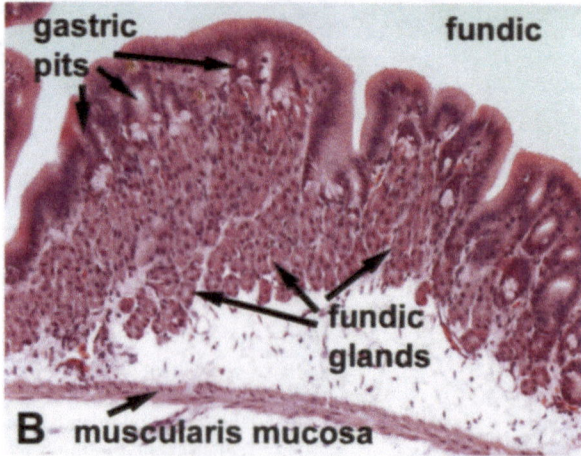

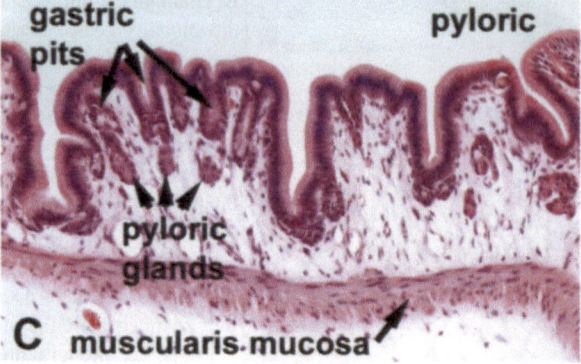

Figure 6. The three major regions of the stomach. **A.** Cardiac stomach. The gastric pits and the cardiac glands are somewhat similar in length, with the glands being somewhat longer. **B.** Fundic stomach. The fundic glands are considerably longer that the gastric pits. **C.** Pyloric stomach. The gastric pits are much longer than the small pyloric glands.

II SMALL INTESTINE

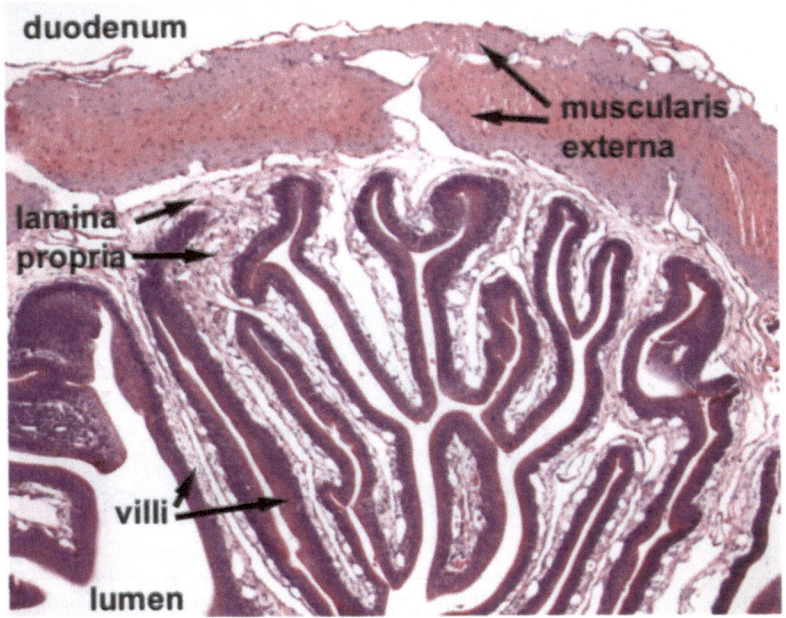

Figure 7. The duodenum of the small intestine. The first region of the small intestine is the duodenum. It has a large muscularis externa. The mucosa evaginates into the lumen as large villi, which serve to increase the absorptive surface area.

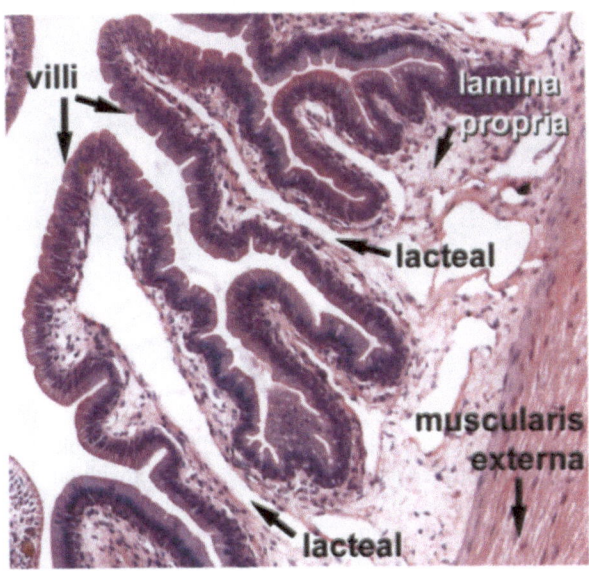

Figure 8. The mucosa of the duodenum. The villi that extend into the lumen are lined by surface epithelium and have a central lacteal that transports absorbed lipids to the circulation. A lamina propria underlies the epithelium, and is in direct contact with the muscularis externa.

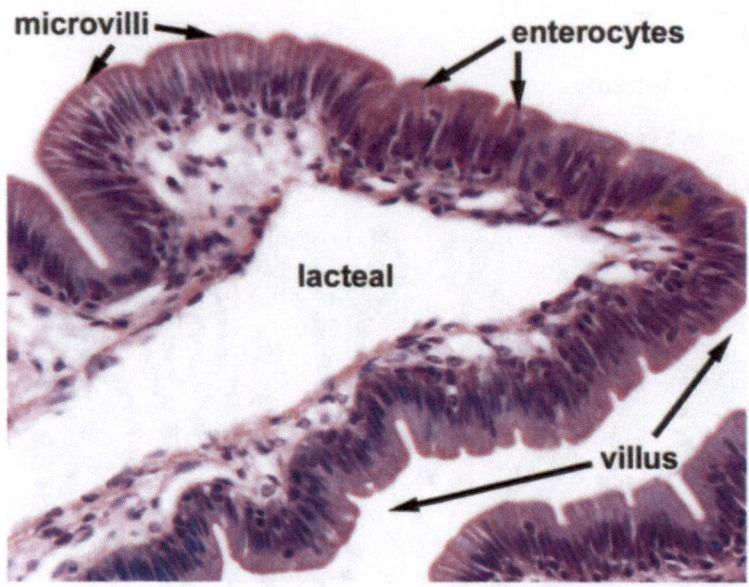

Figure 9. A villus of the duodenum. The surface epithelium of the duodenum is composed of a simple columnar epithelium composed mainly of cells called enterocytes. Mucus-secreting goblet cells are also in the epithelial layer. A large central lacteal is in the lamina propria, which is a lymphatic duct that transports lipids to the circulation.

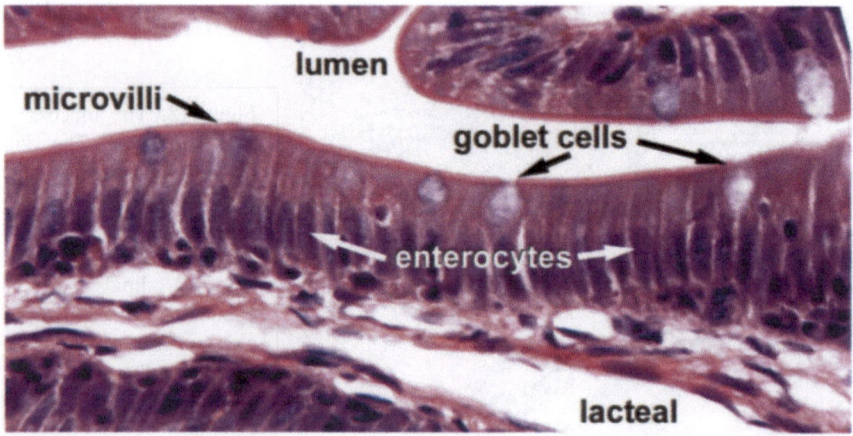

Figure 10. The surface epithelium of the small intestine. The cells of the simple columnar epithelia of the duodenum are called enterocytes. They have microvilli on the apical surface, which serve to increase the absorptive surface area. Mucus-containing goblet cells are also within the epithelium.

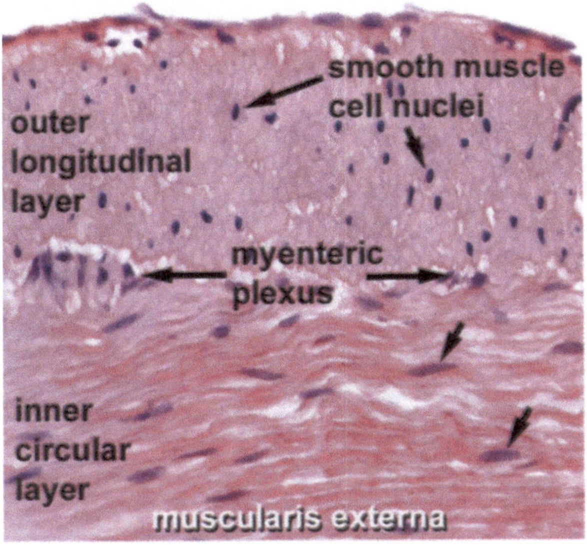

Figure 11. The muscularis externa of the small intestine. The muscularis externa of the duodenum is composed of an inner circular and outer longitudinal layer of smooth muscle. Between the two layers of smooth muscle is the myenteric (or Auerbach's) plexus, which contains the postganglionic parasympathetic neurons that innervate the smooth muscle of the muscularis externa. Covering the outer surface of the muscularis externa is a serosa composed of simple squamous epithelium and some underlying connective tissue.

III   COLON

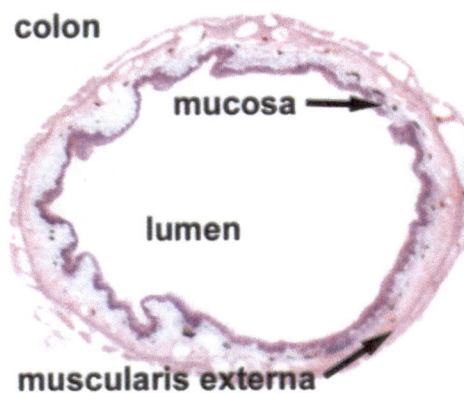

Figure 12. Low magnification image of the colon. A mucosa, comprised of a surface epithelium and lamina propria line the lumen, and a thin muscularis externa forms the outermost layer.

Figure 13. The layers of the colon. The surface epithelium lines the lumen, and has an underlying lamina propria of loose connective tissue. Deep to the lamina propria is a thin muscularis externa.

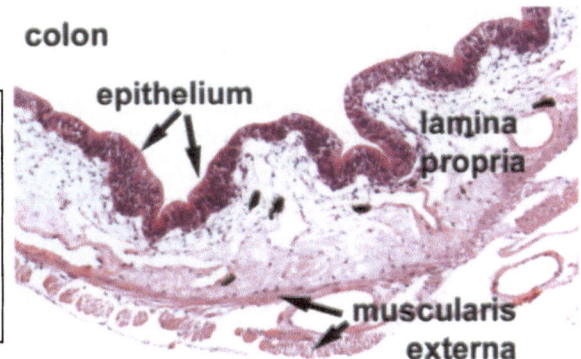

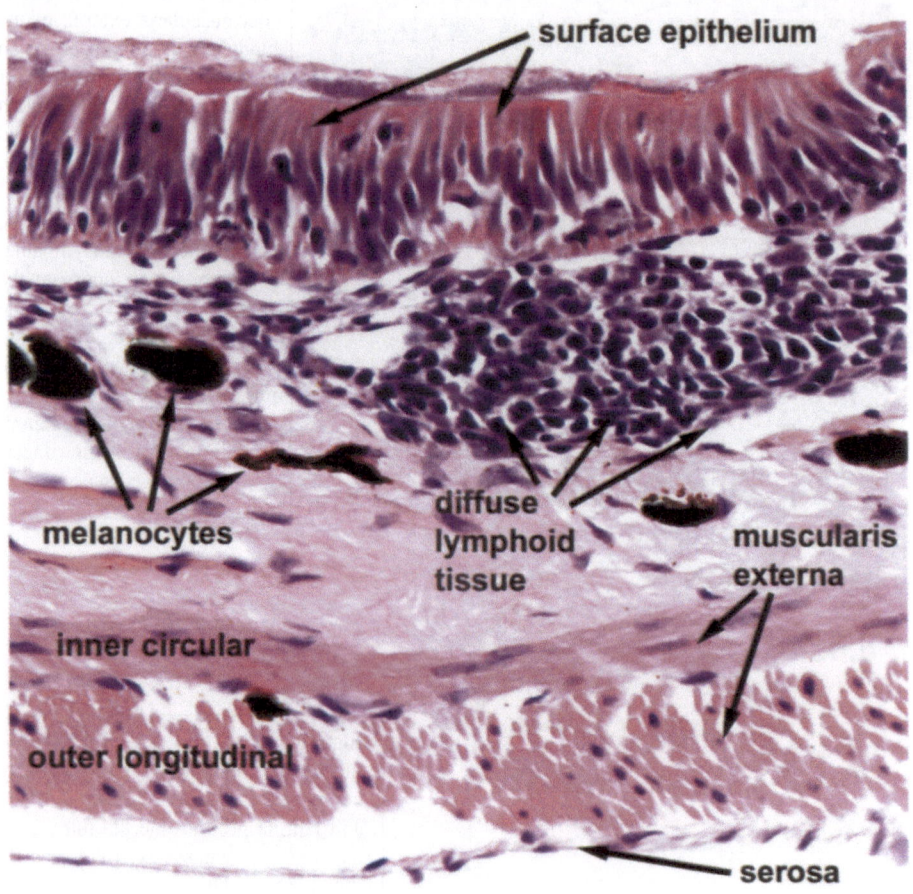

Figure 14. The cells of the colon. The surface epithelium is a simple columnar epithelium with enterocytes and goblet cells. Often overlying the apical surface of the epithelium is ingested debris as seen in this figure. The lamina propria contains many cells types, including fibroblasts and melanocytes. Clusters of diffuse lymphatic tissue, composed mostly of leukocytes and plasma cells, are often present in the lamina propria. The muscularis externa is composed of an inner circular and outer longitudinal layer of smooth muscle. A serosa lines the outer surface of the colon.

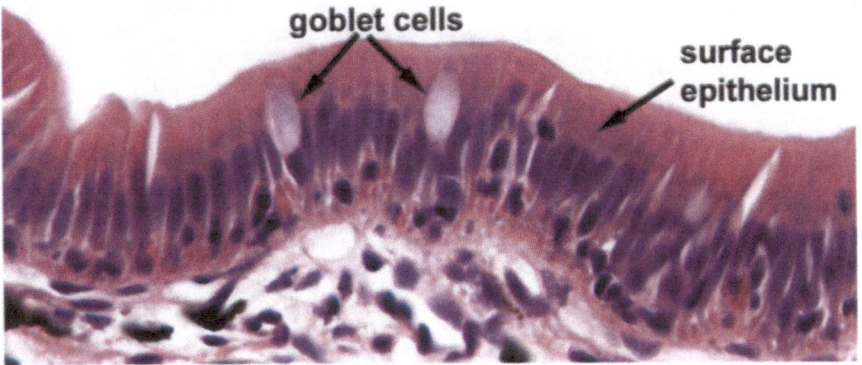

Figure 15. The epithelium of the colon. The simple columnar epithelium of the colon contains mucus-secreting goblet cells. Goblet cells reach the apical surface of the epithelium and release their contents by meroerine secretion (exocytosis).

IV LIVER

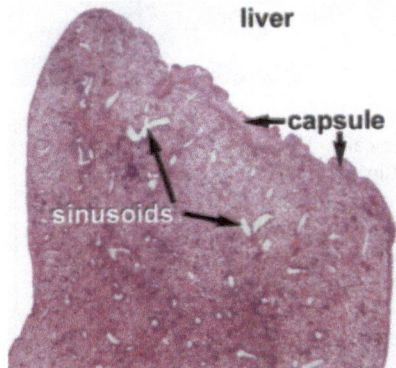

Figure 16. Low magnification image of the liver. The liver is encapsulated by a thick dense connective tissue capsule. Within the parenchyma of the organ are many sinusoids.

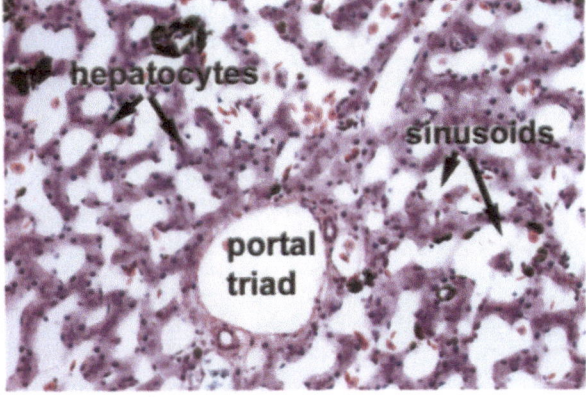

Figure 17. The organization of the liver. The major cell type of the liver are hepatocytes, which are joined together is cords, and are separated by numerous sinusoids. Portal triads are found throughout the liver, and each are composed of an artery, vein and bile duct.

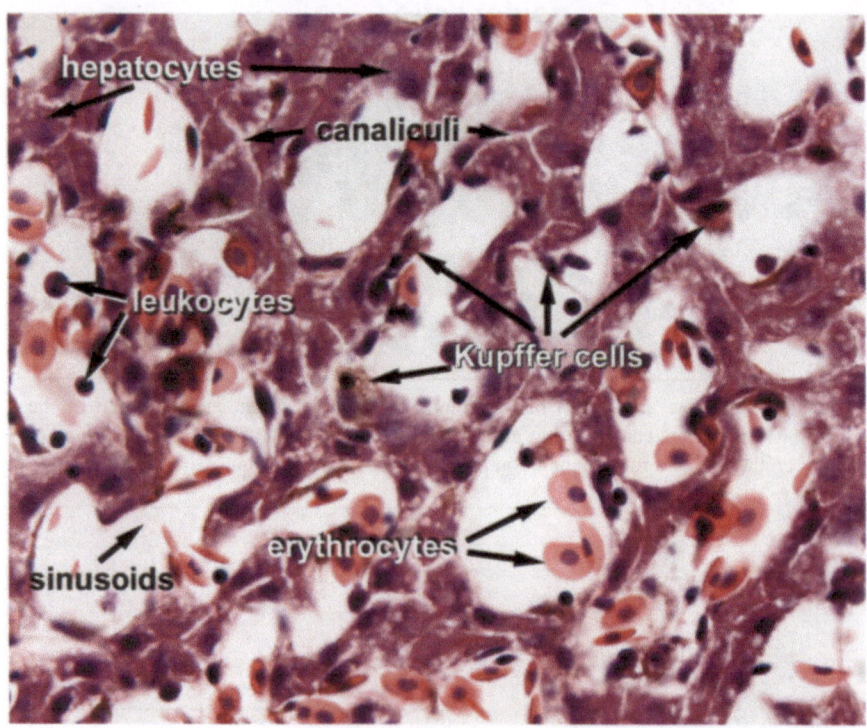

Figure 18. High magnification image of liver cells. The hepatocytes are arranged into cords separated by venous sinusoids. Within the sinusoids are erythrocytes and leukocytes. Also within the sinusoids are Kupffer cells, which are phagocytic, and participate in the removal of senile erythrocytes. The cytoplasm of the hepatocytes has many vacuoles, which are made more visible by the extraction of glycogen during paraffin processing. Canaliculi are small channels between hepatocytes, which transport bile made by the hepatocytes to the bile ducts of the portal triads.

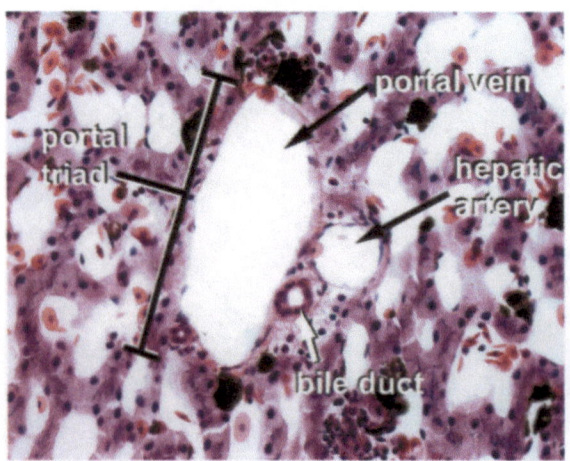

Figure 19. Portal triads of the liver. Portal triads are composed of a 1) portal vein, which brings venous blood from the gut to the liver for detoxification of poisonous compounds, 2) hepatic artery, which transports arterial blood to the hepatocytes, and 3) bile duct, which transports bile from the hepatocyte canaliculi to the common hepatic duct and then to the gall bladder.

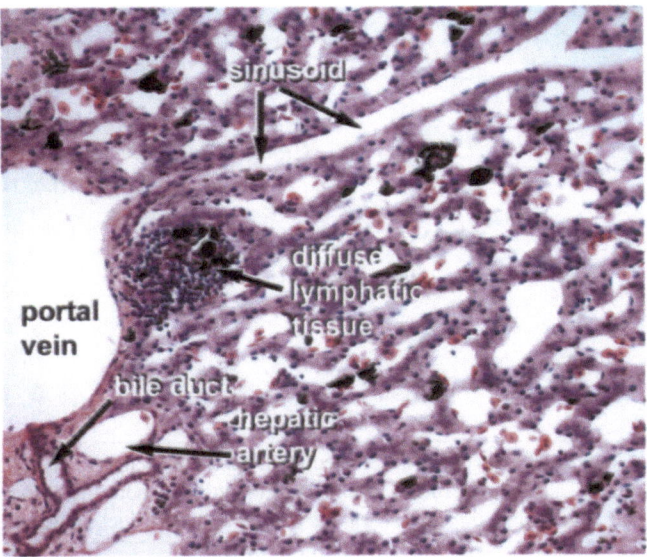

Figure 20. Sinusoids of the liver. The branches of the portal vein that are present in the portal triads send out sinusoidal branches between the cords of hepatocytes. These sinusoids deliver blood directly from the gut to the hepatocytes for detoxification. Note the presence of diffuse lymphatic tissue near the portal triad.

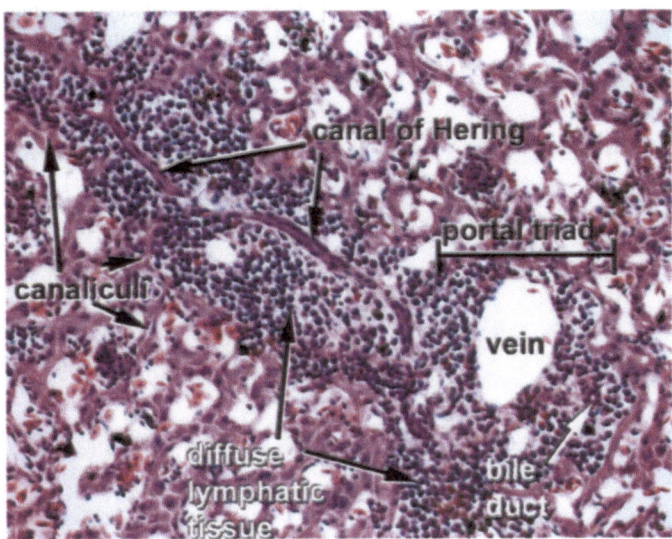

Figure 21. Route of bile flow in the liver. Bile that is produced in the hepatocytes is secreted into canaliculi, which are small channels between hepatocytes. Bile from the canaliculi drains into small ducts called canals of Hering, which transport the bile to the bile duct in the portal triad.

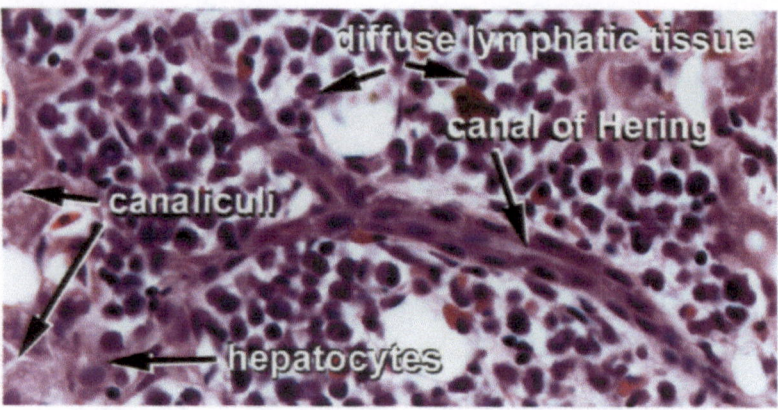

Figure 22. Canals of Hering. Bile is secreted from hepatocytes into the canaliculi, which merge to form the canals of Hering. The canals of Hering transport bile to the bile duct in the portal triad. Bile is then transported via the common hepatic duct to the gall bladder.

V GALL BLADDER

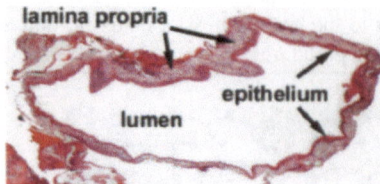

Figure 23. Low magnification image of the gall bladder. The lumen is lined by a simple columnar epithelium with an underlying lamina propria.

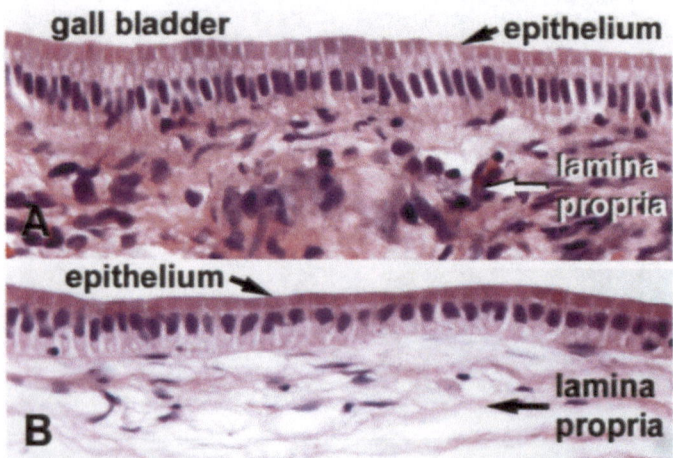

Figure 24. Epithelium of the gall bladder. The lumen of the gall bladder is lined by a simple columnar epithelium. Deep to the epithelium is a loose connective tissue lamina propria.

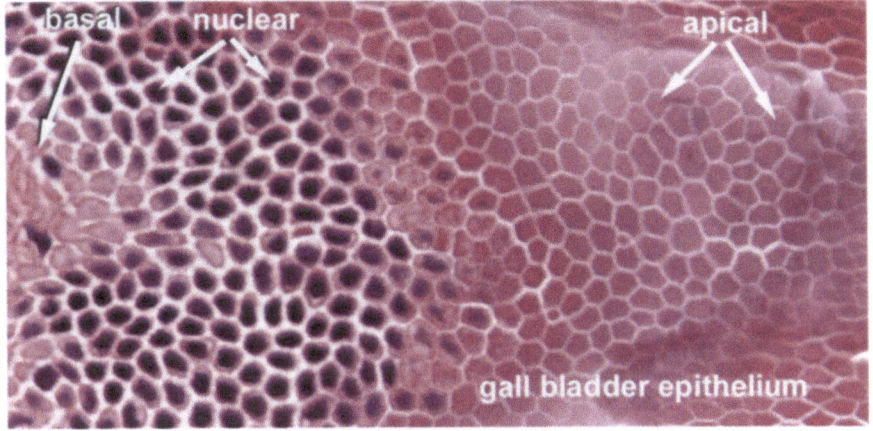

Figure 25. Oblique orientation of the epithelium of the gall bladder. The most apical region of the simple columnar epithelium shows a hexagonal appearance. Further down in the cell is the nucleus, followed by the basal portion of the cell.

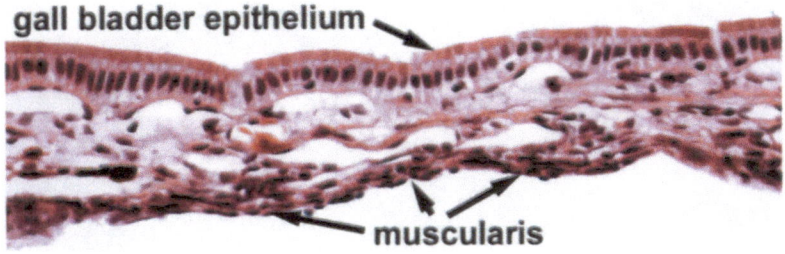

Figure 26. Muscularis of the gall bladder. Deep to the lamina propria of the gall bladder is a thin layer of smooth muscle that comprises the muscularis. External to the muscularis is the adventitia.

VI PANCREAS

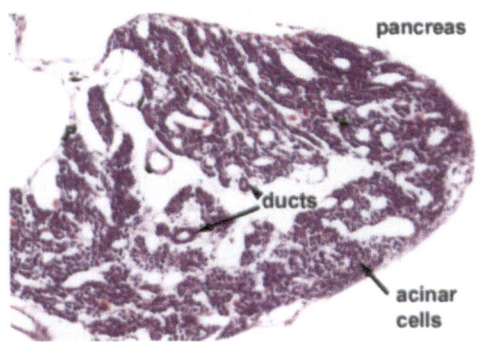

Figure 27. Low magnification image of the pancreas. The majority of the pancreas is comprised of exocrine pancreatic acinar cells that expel their contents through a series of ducts.

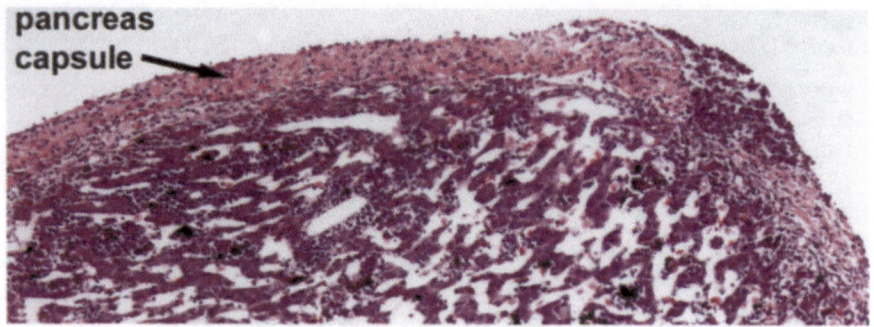

Figure 28. Capsule of the pancreas. The pancreas is encapsulated by a layer of dense connective tissue.

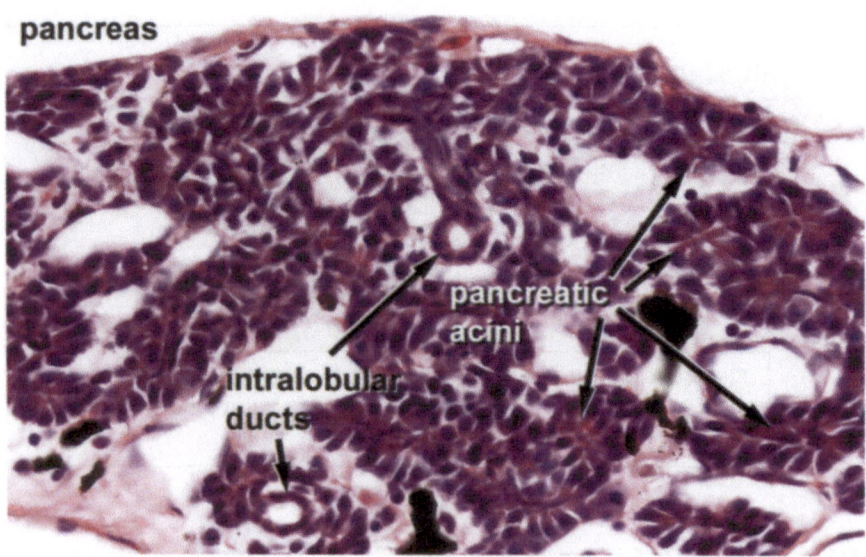

Figure 29. Exocrine portion of the pancreas. The pancreatic acinar cells form pancreatic acini, which are clusters of acinar cells. Their apical surfaces are in contact with a small lumen that is continuous with intercalated ducts that drain into larger intralobular ducts. The acinar cells secrete digestive enzyme into the lumen, which are eventually secreted into the duodenum.

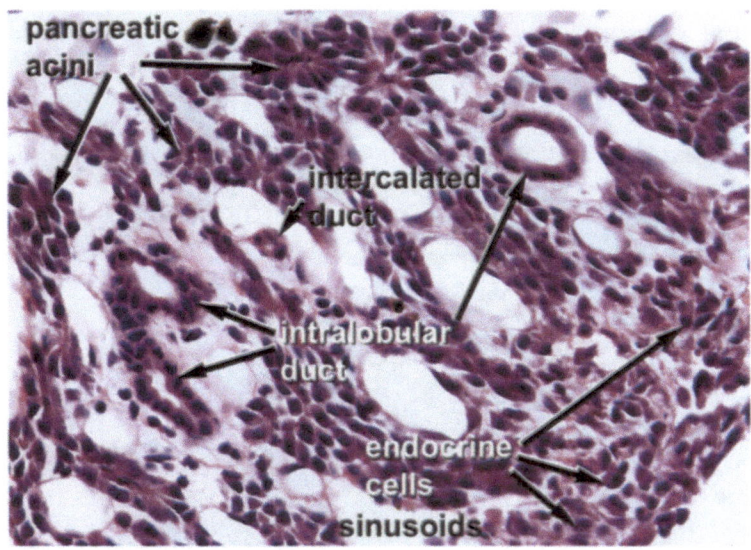

Figure 30. Cytoarchitecture of the pancreas. Intercalated ducts arise directly from the pacreatic acini, which drain into the larger intralobular ducts. The intercalated ducts secrete bicarbonate and water into the lumen, which helps to neutralize the low pH in the chyme that enters the duodenum from the stomach. The hormone-secreting endocrine cells of the pancreas are arranged in clusters of varying sizes.

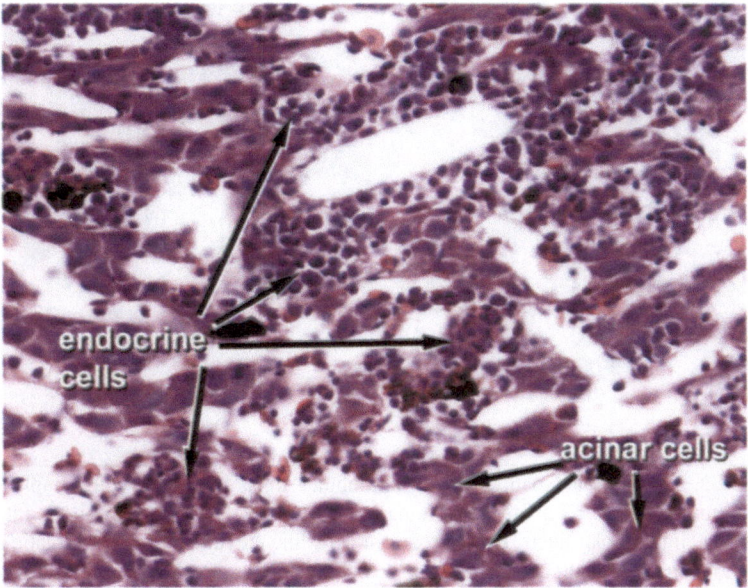

Figure 31. Endocrine cells of the pancreas. The endocrine cells are dispersed throughout the pancreas in irregular clusters. They secrete insulin, glucagon, and other hormones that help to regulate glucose metabolism.

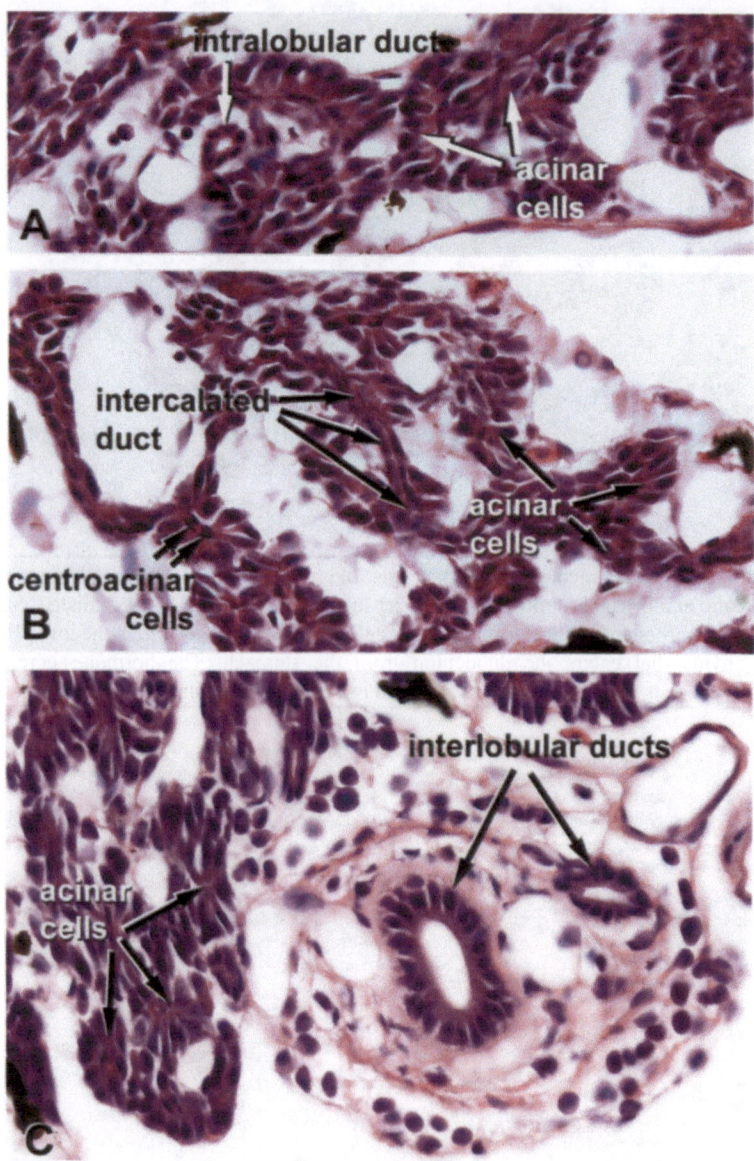

Figure 32. Ducts of the pancreas. **A.** Intralobular ducts are lined with simple cuboidal epithelium and have a relatively large lumen. They transport secretions from the intercalated ducts to the interlobular ducts. **B.** Centroacinar cells line the lumen of the pancreatic acini and are continuous with intercalated ducts. Intercalated ducts are lined with a low cuboidal epithelium and have a very small lumen. They transport secretions from the pancreatic acinar cells to the intralobular ducts and also secrete bicarbonate and water. **C.** Interlobular ducts are located in connective tissue septa between lobules and receive secretions from the intralobular ducts. They are lined with a simple or stratified cuboidal epithelium.

Chapter 5

Respiratory System

The upper respiratory system warms, moistens, and filters the inhaled air and provides a passageway to the lungs where oxygen is delivered to erythrocytes in the circulation, and carbon dioxide is passed from the erythrocytes into the air chambers to be exhaled from the body.

I NASAL CAVITY

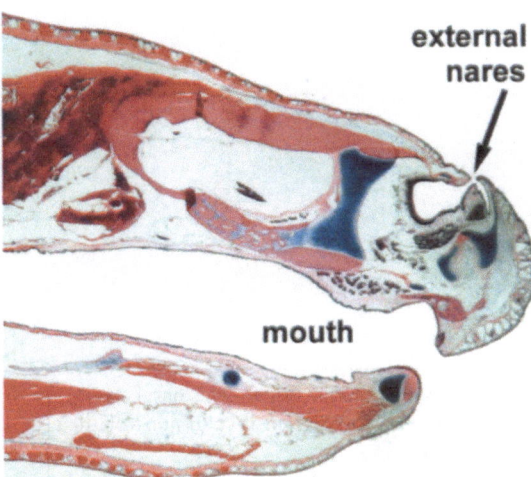

Figure 1. Low magnification of sagittal section of *Xenopus* head. The respiratory tract begins at the external nares.

Figure 2. Chambers of the external nares. The external nares constricts to specifically admit air in the one chamber (1) that leads to the nasal cavity when the nares are above the water level. Below the water level, the nares constricts to admit water into a second chamber (2) lined with olfactory epithelium, thus enabling the frog to detect water-borne odorants.

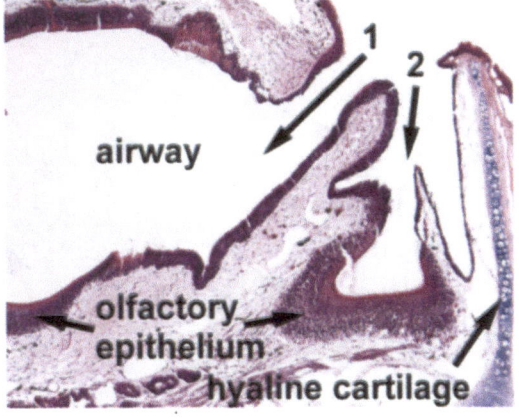

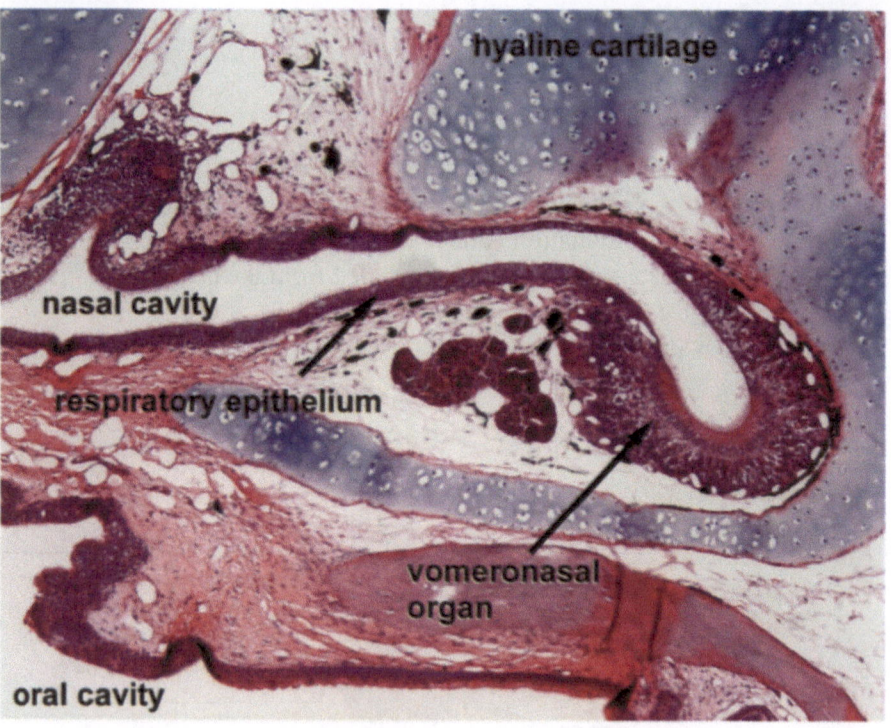

Figure 3. Sagittal section of the vomeronasal organ. The nasal cavity delivers fluid to the vomeronasal organ, which is sensitive to non-volatile substance such as pheromones.

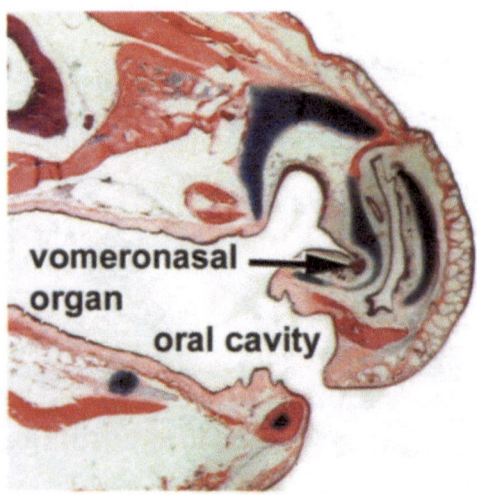

Figure 4. Low magnification of a sagittal section of a frog head showing the location of the vomeronasal organ. The vomeronasal organ is located just above the oral cavity.

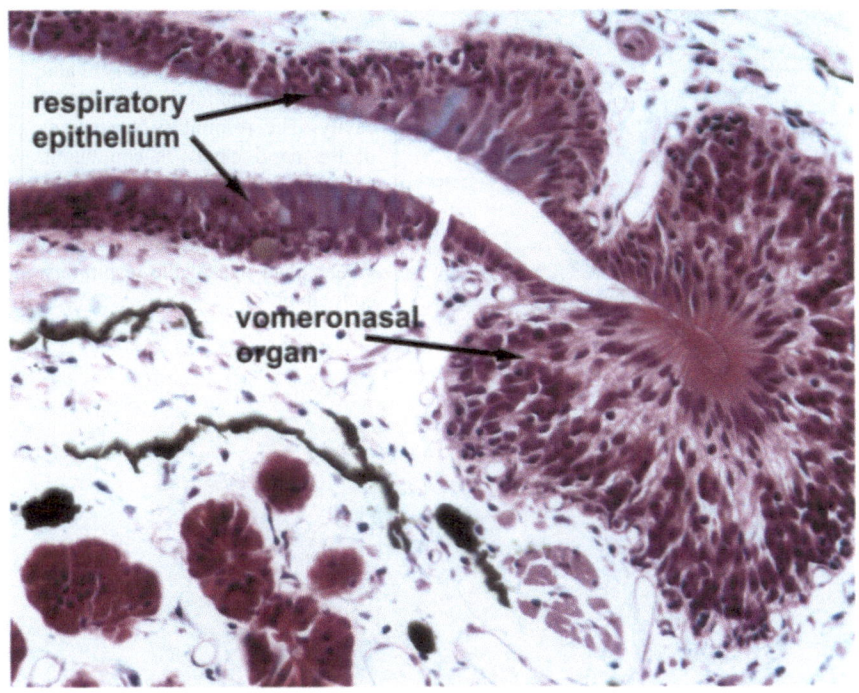

Figure 5. High magnification of the vomeronasal organ. The sensory epithelium is continuous with the respiratory epithelium of the nasal cavity.

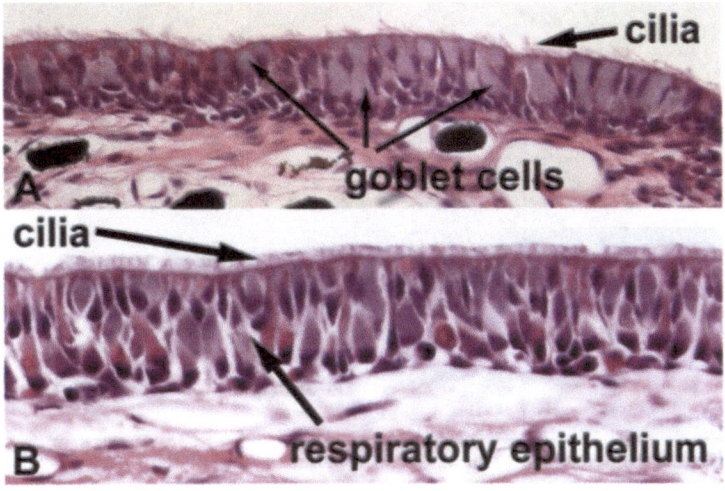

Figure 6. Respiratory epithelium of the nasal cavity. **A.** Some regions of the nasal cavity are lined with respiratory epithelium with many goblet cells. **B.** Respiratory epithelium lining most of the nasal cavity is highly ciliated and has fewer goblet cells.

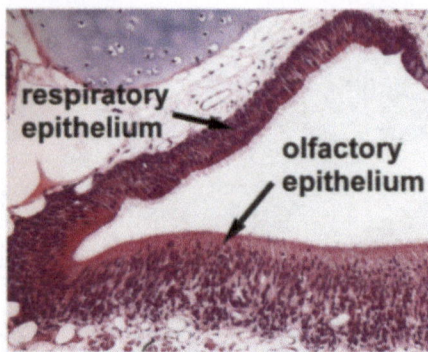

Figure 7. Respiratory and olfactory epithelium of the nasal cavity. The respiratory epithelium of the nasal cavity is continuous with the sensory olfactory epithelium. Note that the thickness of the olfactory epithelium is much greater than that of the respiratory epithelium.

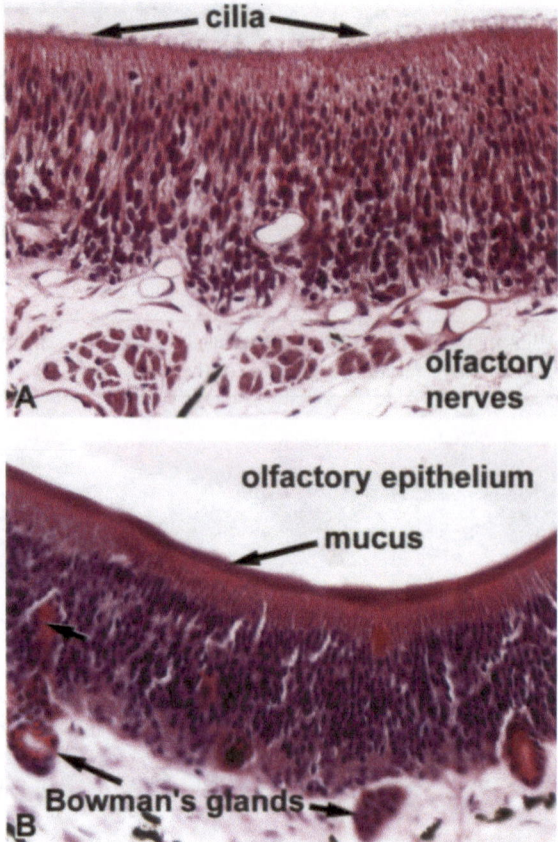

Figure 8. Olfactory epithelium. **A.** The olfactory epithelium is a pseudostratified columnar epithelium with sensory cilia on the apical surface. Note the olfactory nerve bundles in the underlying connective tissue. **B.** In some areas of the nasal cavity, the apical surface of the olfactory epithelium is covered with a layer of mucus, which entraps air-borne odorants. The mucus is produced by Bowman's glands in the underlying connective tissues, which send ducts to the epithelial surface.

II TRACHEA

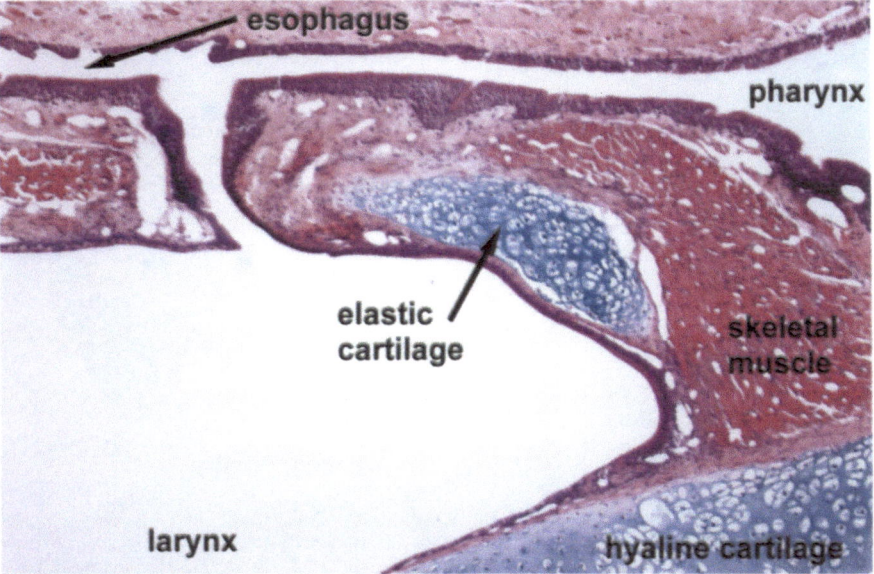

Figure 9. Epiglottis. Between the pharynx and trachea is the epiglottis, which prevents ingested materials from entering the airway of the trachea. Skeletal and smooth muscle surround a core of elastic cartilage. Closure of the epiglottis forces ingested materials into the esophagus, which is posterior to the trachea.

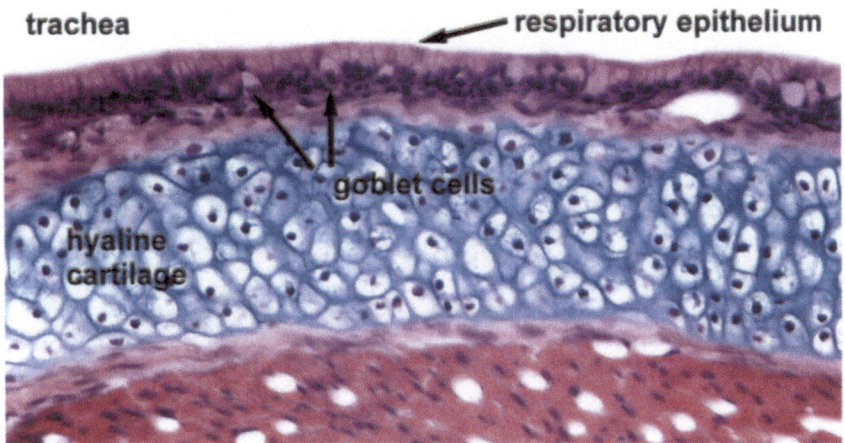

Figure 10. Tissues of the trachea. The luminal surface of the trachea is lined with ciliated respiratory epithelium with goblet cells. Deep to the epithelium is a ring of hyaline cartilage, which provides structural support to prevent collapse of the airway.

III LUNGS

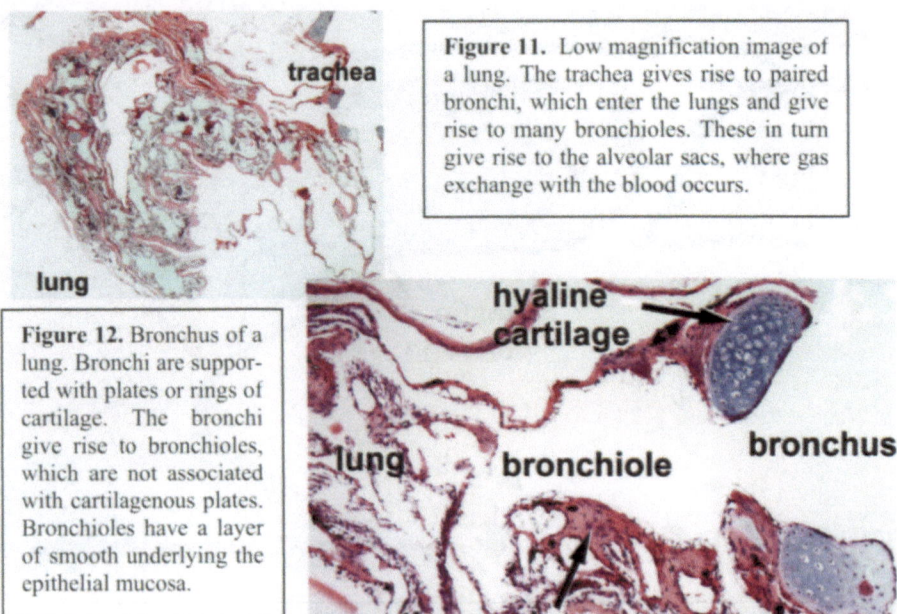

Figure 11. Low magnification image of a lung. The trachea gives rise to paired bronchi, which enter the lungs and give rise to many bronchioles. These in turn give rise to the alveolar sacs, where gas exchange with the blood occurs.

Figure 12. Bronchus of a lung. Bronchi are supported with plates or rings of cartilage. The bronchi give rise to bronchioles, which are not associated with cartilagenous plates. Bronchioles have a layer of smooth underlying the epithelial mucosa.

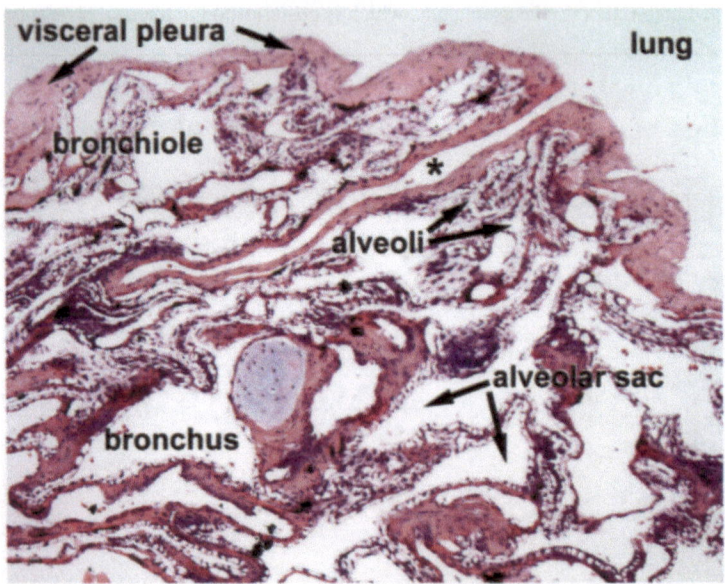

Figure 13. Low magnification image of lung tissues. A visceral pleura forms the outer surface of the lung. Note the invaginations of the lung surface (*). Bronchi are identified by the presence of the basophilic hyaline cartilage. Bronchioles are identified by the presence of smooth muscle and lack of alveoli.

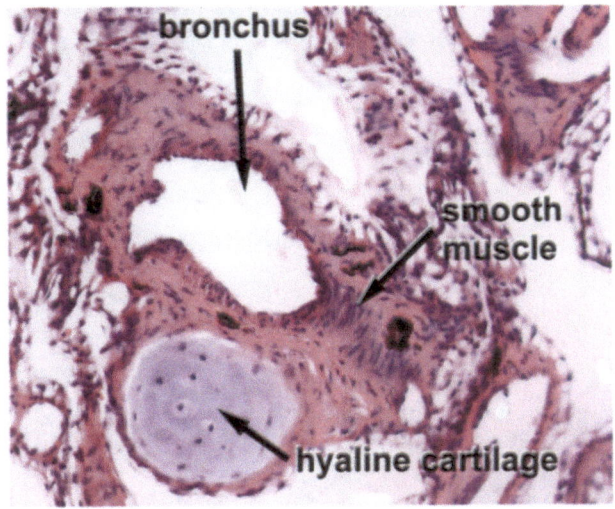

Figure 14. Bronchus of the lung. Bronchi are identified by the presence of hyaline cartilage and smooth muscle below the epithelial layer. Note the orientation of the smooth muscle cells (similar to the orientation of smooth muscle cells that occurs in arteries) in this oblique section.

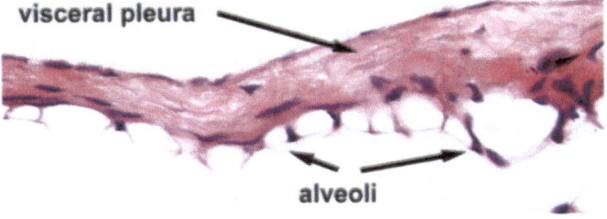

Figure 15. High magnification of the visceral pleura. The outer surface is lined with a layer of simple squamous epithelium. The inner surface is often lined with alveolar capillaries where gas exchange occurs.

Figure 16. Alveolar sacs of the lung. The bronchioles, which are not lined with alveolar capillaries, give rise to alveolar sacs. Alveolar sacs are blind-ended pouches lined with alveolar capillaries and pneumocytes. This is the location where gas exchange occurs with the erythrocytes.

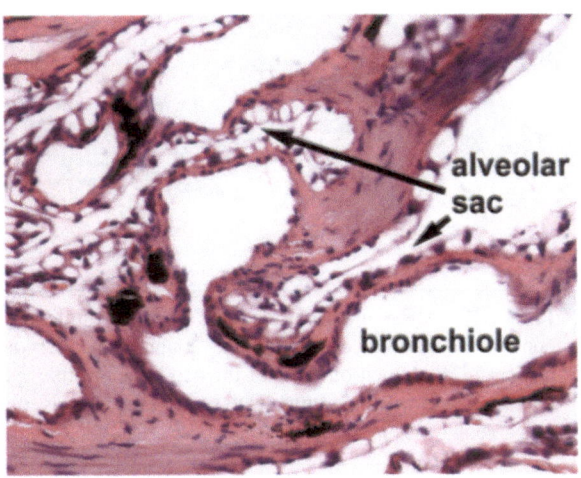

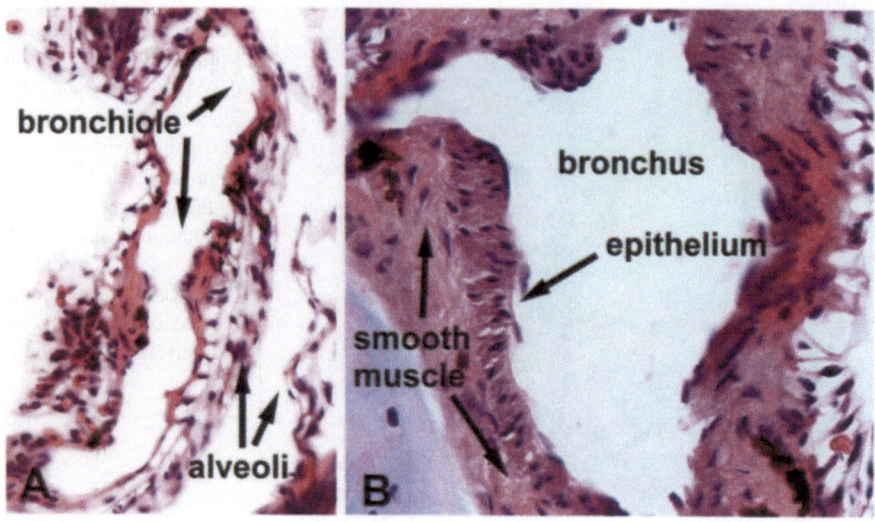

Figure 17. Airways of the lung. Depending on the level of branching, the bronchus is lined with columnar, cuboidal, or simple squamous epithelium, with columnar epithelium at the highest levels of branching, and simple squamous at the lowest level (just before giving rise to bronchioles). Bronchioles are lined with a layer of simple squamous epithelium. Smooth muscle underlies the epithelium in bronchi and bronchioles.

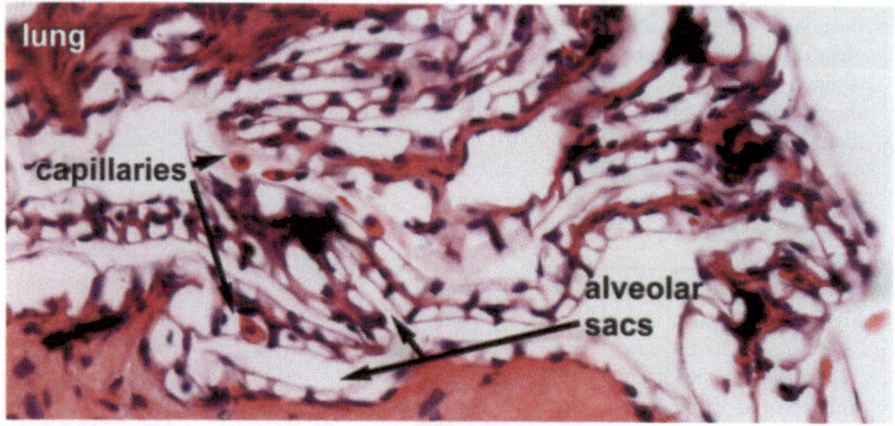

Figure 18. Alveolar sacs of the lung. The bronchioles give rise to alveolar sacs, which contain the alveolar capillaries and pneumocytes. Erythrocytes (red blood cells) are present in the lumen of some of the capillaries in this specimen. Note that an alveolar duct (passageway to the immediate right of the two alveolar sacs indicated) gives rise to two alveolar sacs in the section.

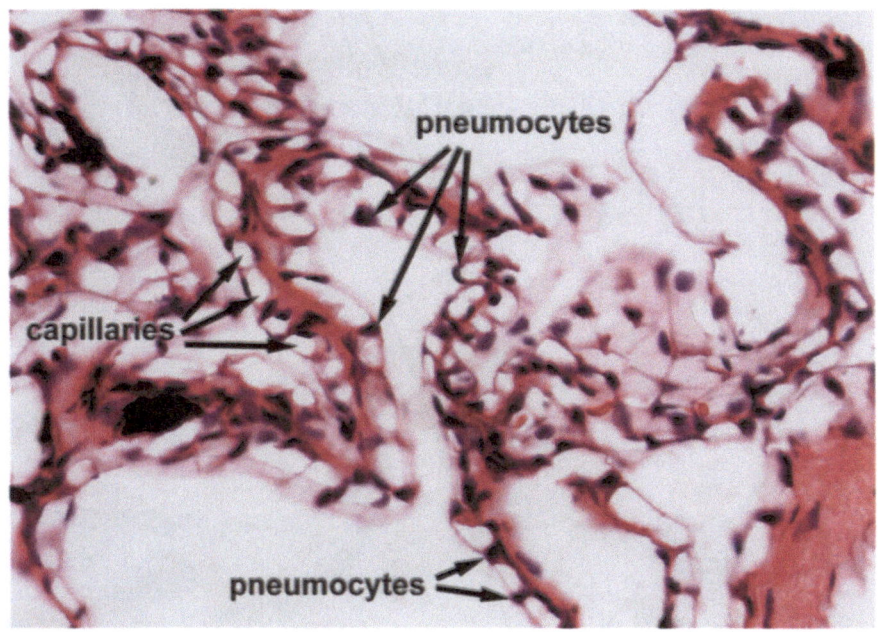

Figure 19. Alveolar sacs of the lung. The cell somas of the pneumocytes are wedged between the cell membranes of the capillaries. Note the alveolar sac that contains the capillaries and pneumocytes indicated, and the alveolar duct from which it arises.

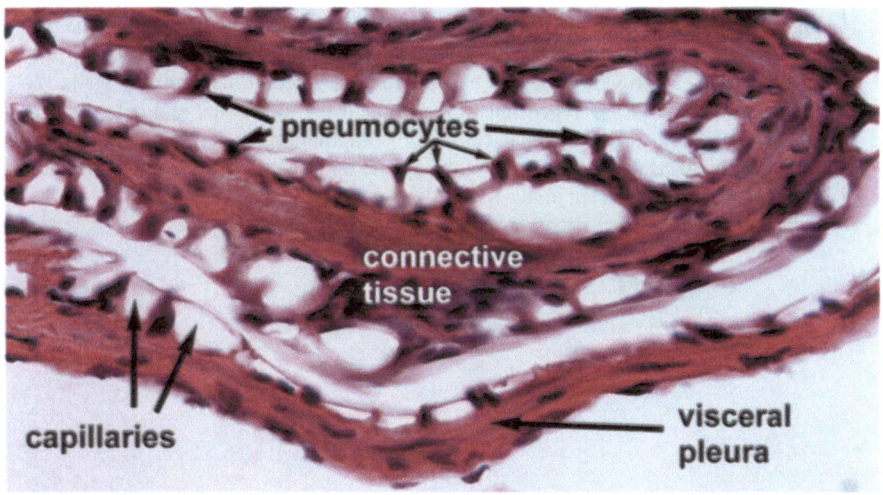

Figure 20. Pneumocytes and capillaries of the lung. The pneumocyte cell bodies are located between adjacent alveolar capillaries. Together, the pneumocytes and capillary endothelial cells comprise the air-blood barrier.

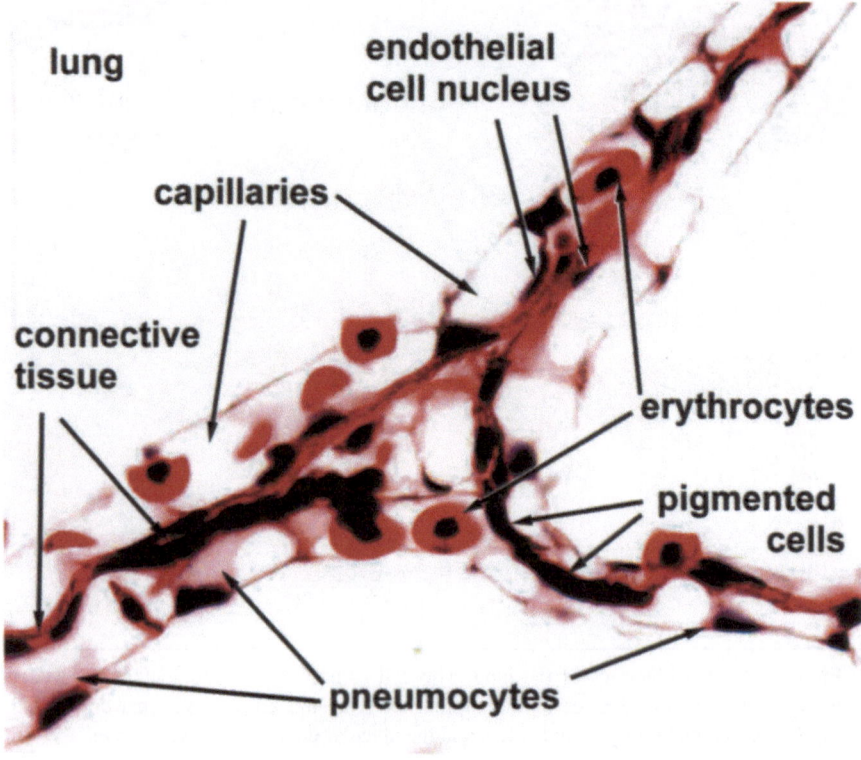

Figure 21. Pneumocytes and capillaries of the lung. The pneumocyte cell bodies are located between adjacent alveolar capillaries, and their thin cytoplasmic sheets overly the capillaries to provide a barrier between the airway and the capillary walls. To reach the erythrocytes, oxygen must pass through the membrane and cytoplasm of the pneumocytes and endothelial cells of the capillaries. Underlying the capillary endothelial cells is their basement membrane and underlying connective tissue. Many pigmented cells are present in the connective tissue of the lung.

Chapter 6

Urinary System

The urinary system is comprised of the kidneys, ureter, urinary bladder, and urethra. Blood is filtered in the kidneys, and the waste products and water pass from the kidneys into the lumen of the ureter. The ureter empties into the urinary bladder which temporarily stores the urine until it is expelled through the urethra.

I KIDNEY

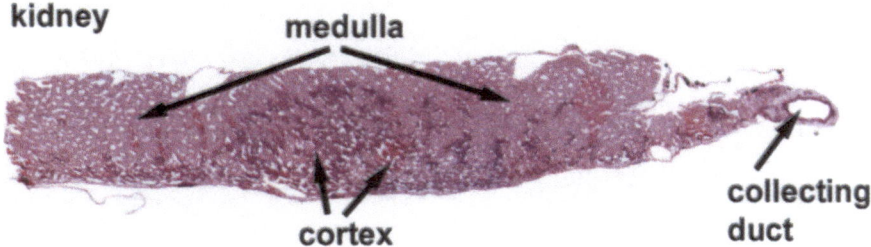

Figure 1. Low magnification image of the *Xenopus* kidney. The kidney has an elongate shape, and a distinct medulla and cortex are present. Large collecting ducts are located toward the end of the organ.

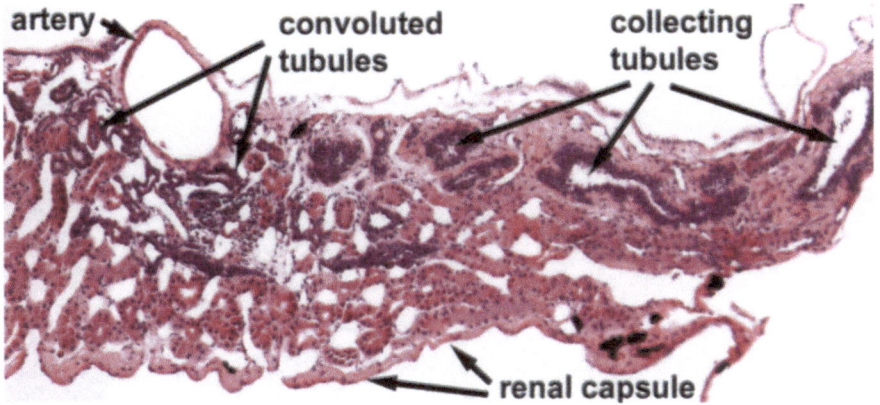

Figure 2. Organization of the kidney. The convoluted tubules are located in the cortex, and the more eosinophilic medulla contains collecting tubules that drain into progressively larger tubules to exit the kidney.

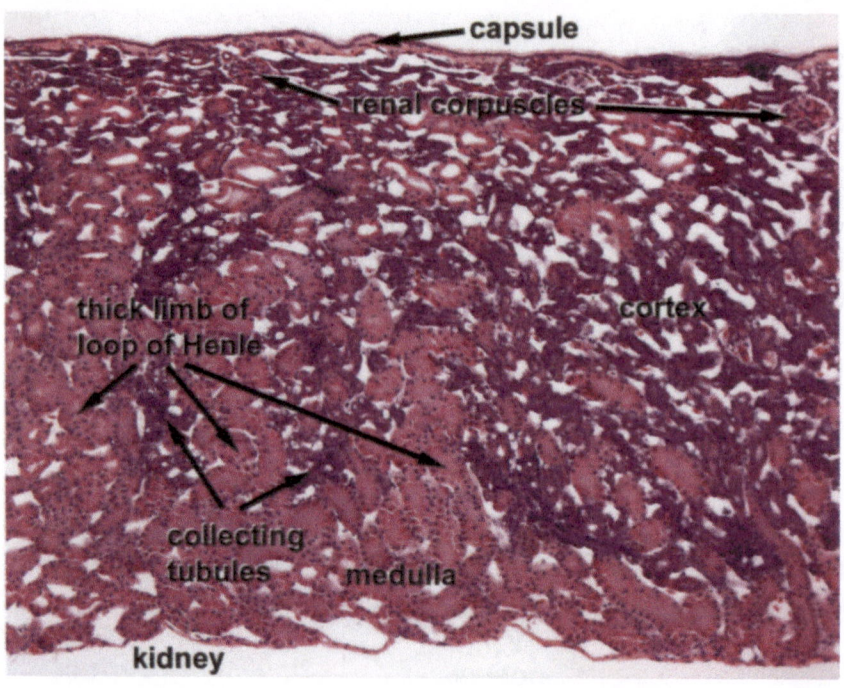

Figure 3. Distribution of structures in the cortex and medulla. The renal corpuscles and convoluted tubules are located in the cortex, and the medulla contains mostly the loop of Henle and collecting tubules.

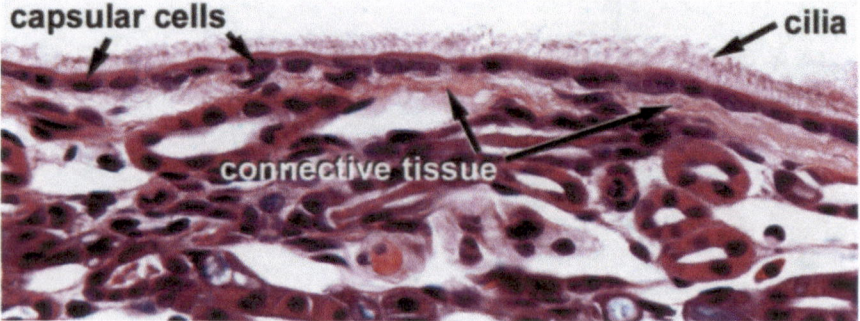

Figure 4. Structure of the renal capsule. There are two major components of the renal capsule; a layer of simple cuboidal epithelium, and an underlying layer of dense connective tissue. Note the presence of cilia on the apical surface of the epithelium.

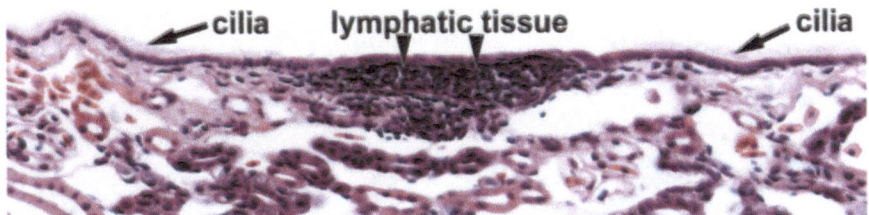

Figure 5. Diffuse lymphatic tissue in the kidney. Clusters of lymphocytes occur throughout the kidney, often associated with collecting tubules. Lymphatic tissue is also located just under the renal capsule. Note the lack of cilia on the cuboidal capsular cells that overly the lymphatic tissue.

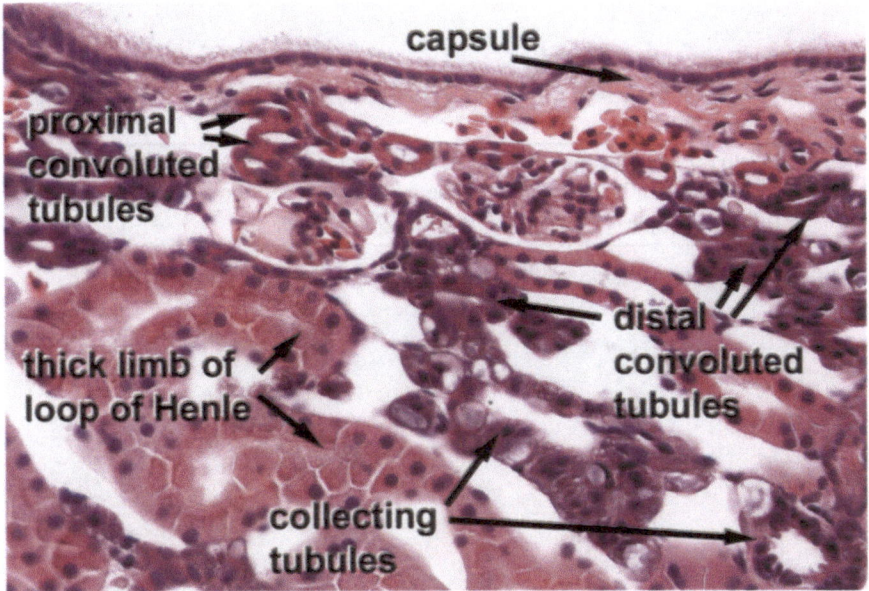

Figure 6. Major tubular structures of the cortex. Underlying the renal capsule are the major tubular structures of the cortex. The proximal convoluted tubules are eosinophilic cuboidal epithelium and are often found near the renal corpuscles. They have microvilli on their apical surface. Also near the renal corpuscles are the distal convoluted tubules which are cuboidal epithelium and have a basophilic appearance. These cells are often larger than the proximal tubule cells. The thick limb of the loop of Henle is continuous with the proximal tubules and has a similar eosinophilic staining quality. The cuboidal cells are larger and also have microvilli on their apical surface. Collecting tubules are quite basophilic, and are composed of two cells types; cuboidal epithelium and mucus-containing flask cells.

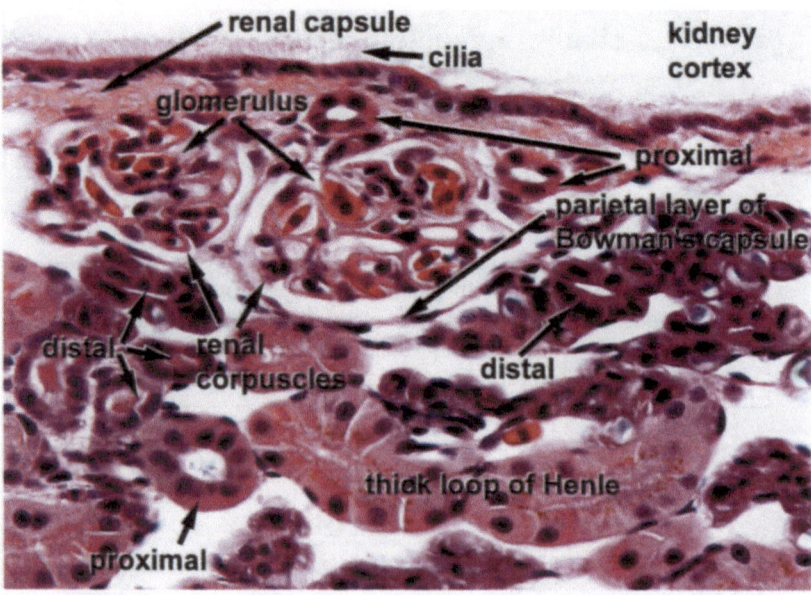

Figure 7. Renal corpuscles and other cortical structures. Most renal corpuscles are located very near the renal capsule. At the center of the renal corpuscle is the capillary-filled glomerulus.

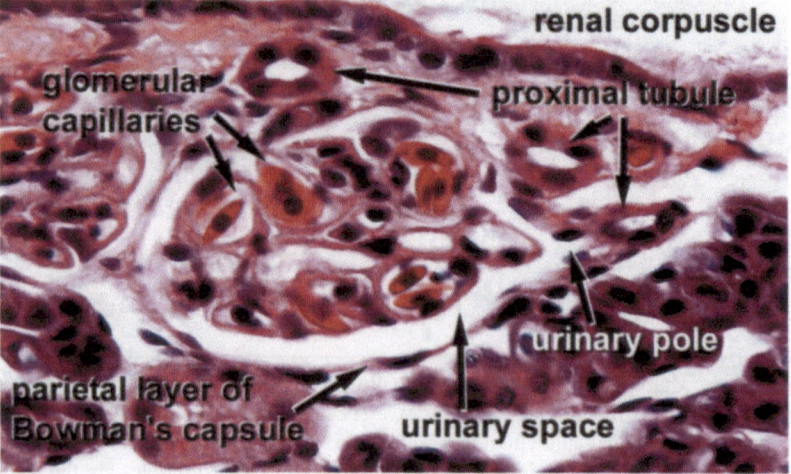

Figure 8. High magnification of the renal corpuscle. The glomerulus is composed primarily of capillaries and the podocytes that comprise the visceral layer of Bowman's capsule. Bowman's capsule consists of the parietal and visceral layers and the urinary space. The parietal layer is composed of simple squamous epithelium. Ultrafiltrate passes through the fenestrated capillaries and between the podocyte cell processes to reach the urinary space from which it then passes into the urinary pole via the proximal convoluted tubule.

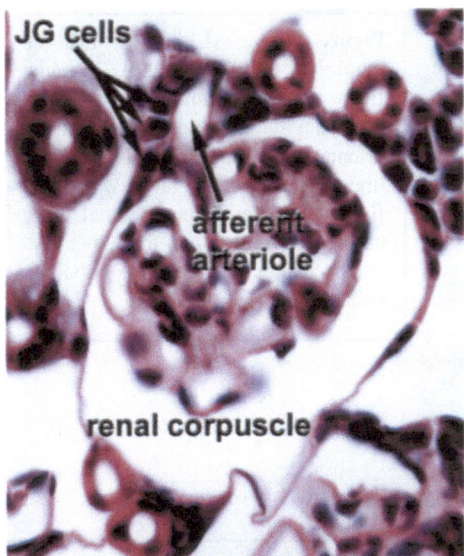

Figure 9. High magnification of a renal corpuscle. The afferent and efferent arterioles enter and leave the glomerulus at the vascular pole. Some smooth muscle cells of the afferent arteriole are specialized cells called juxtaglomerular cells (JG cells) and secrete renin in response to signals from the nearby macula densa of the distal convoluted tubule.

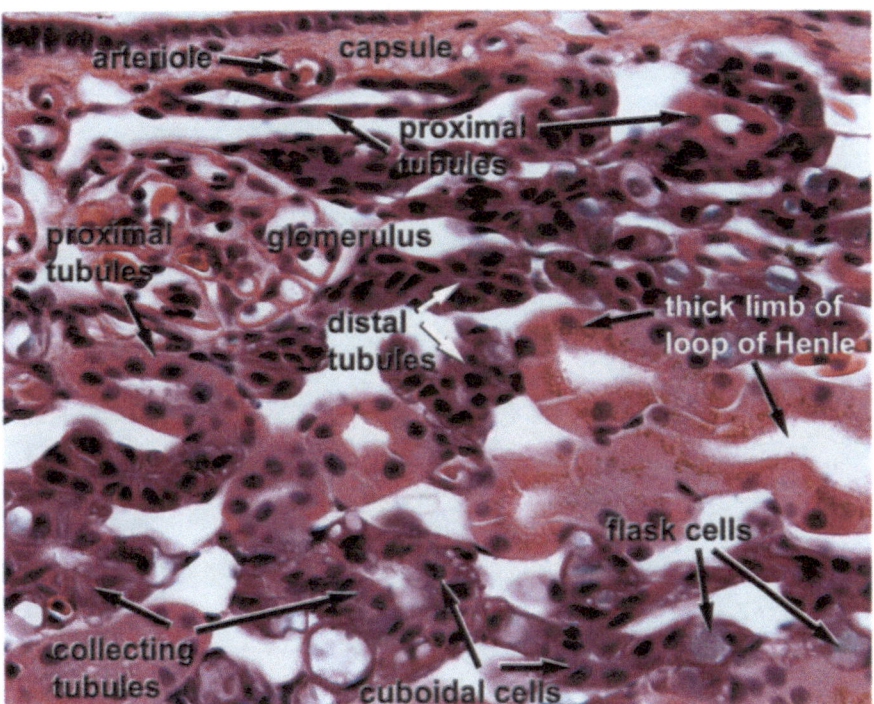

Figure 10. High magnification of renal cortex. In addition to proximal and distal convoluted tubules and renal corpuscles, collecting tubules are also located in the renal cortex (and in the medulla). The cells comprising the collecting tubules are cuboidal epithelial cells and mucus-secreting cells called flask cells.

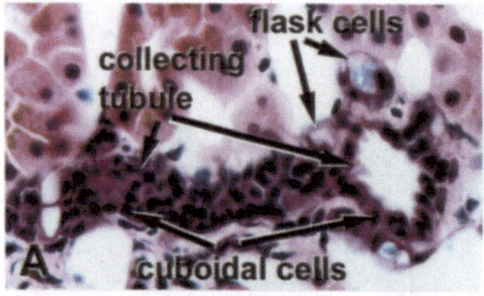

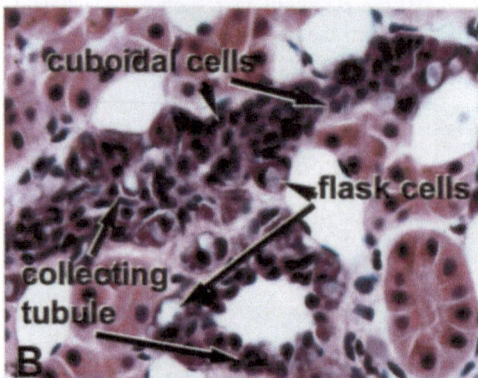

Figure 11. (LEFT) Collecting tubules in the renal medulla. A and B. A collecting tubule is viewed in both longitudinal and cross sectional orientations. Cuboidal epithelial cells line the lumen of the tubules, and the flask cells overlie the cuboidal cells and secrete mucus into the lumen via small channels between the cuboidal cells.

Figure 12. (BELOW) Collecting tubules in the renal medulla. Note that several distal convoluted tubules drain into a single larger collecting tubule (longitudinal orientation).

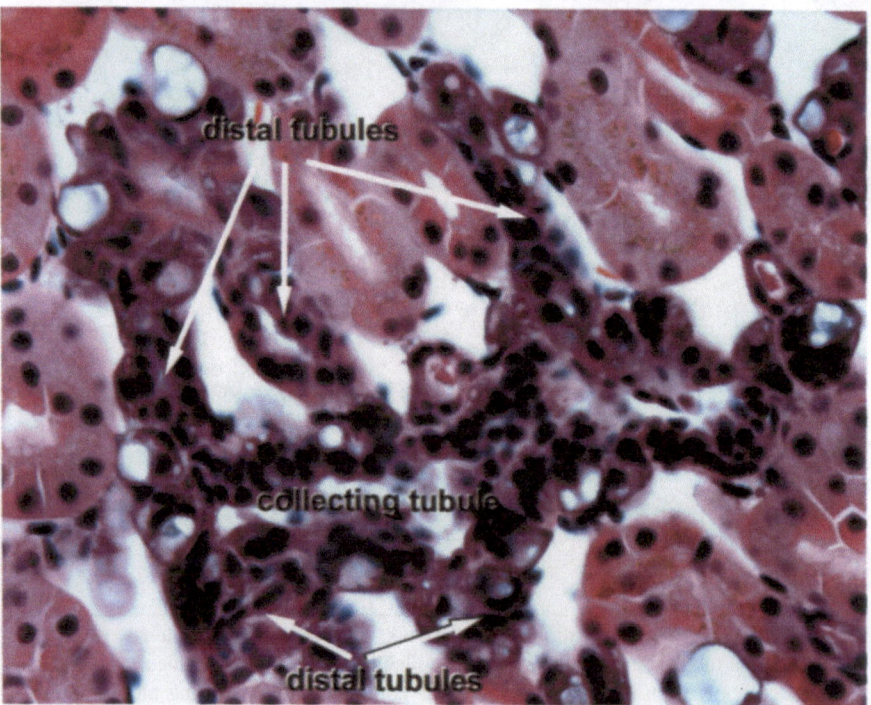

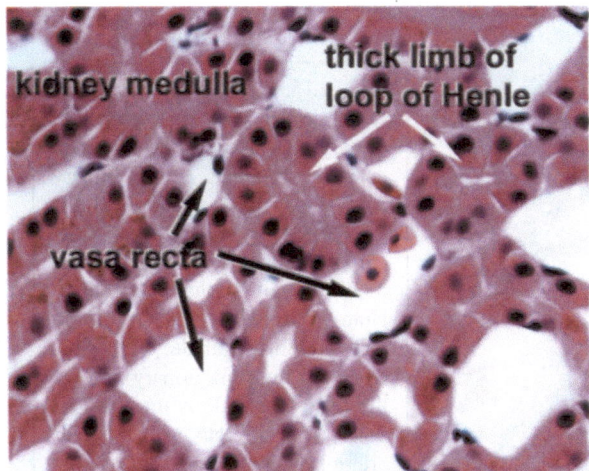

Figure 13. Renal medulla. The major structures in the renal medulla are the thick and thin limbs of the loop of Henle and the vasa recta. The cells of the thick limb of the loop of Henle are large, pick-staining cuboidal epithelial cells. The vasa recta are composed of a layer of endothelium and may have erythrocytes in the lumen.

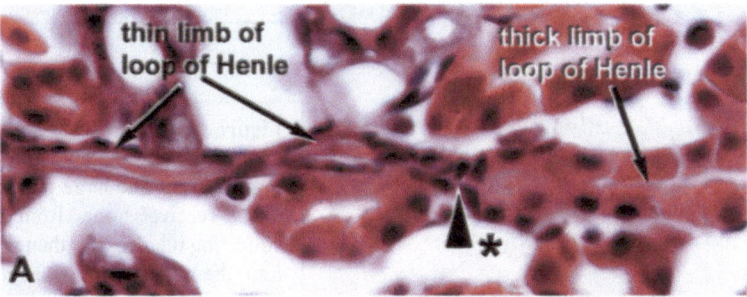

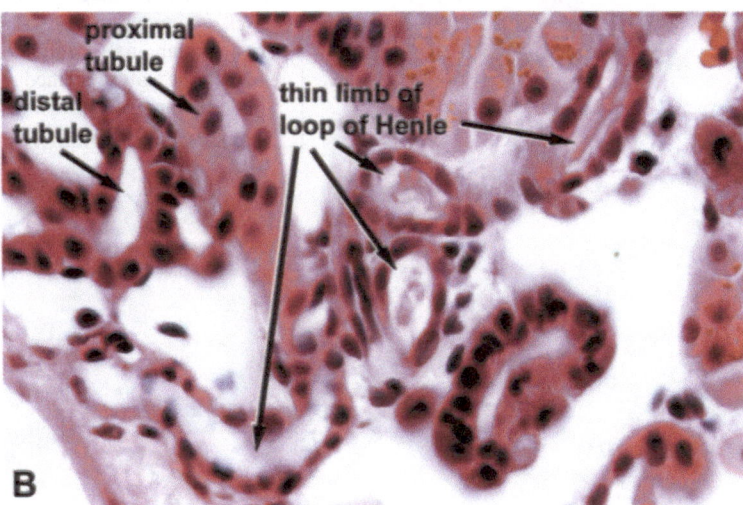

Figure 14. Loop of Henle in the renal medulla. **A.** Longitudinal section of the loop of Henle showing the transition (*) from the cuboidal cells of the thick limb to the squamous cells of the thin limb. **B.** Thin limb of the loop of Henle is shown in several orientations.

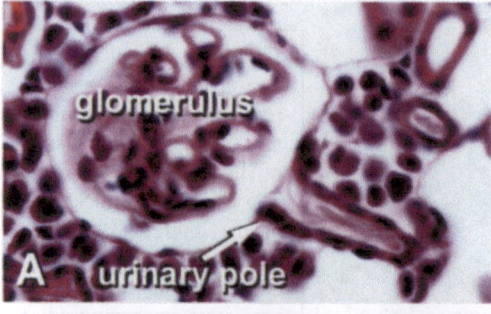

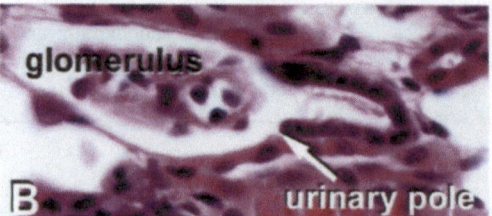

Figure 15. Renal corpuscles in the renal cortex. **A.** Ultrafiltrate in urinary space enters the proximal convoluted tubule at the urinary pole of the renal corpuscle. **B.** At the urinary pole of the renal corpuscle, the proximal convoluted tubule receives the ultrafiltrate. Note the presence of microvilli on the proximal tubule cells. **C.** Opposite from the urinary pole of the renal corpuscle is the vascular pole. The afferent and efferent arterioles enter and leave the glomerulus at the vascular pole.

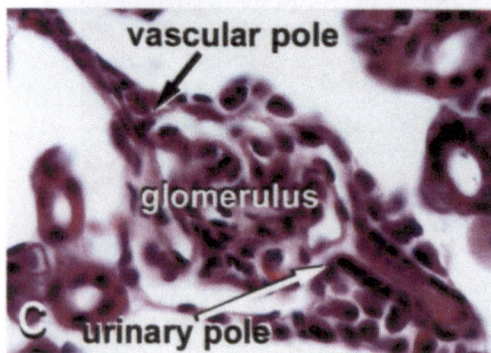

Figure 16. (BELOW) Papillary ducts drain into the lumen of the minor calyx. The large papillary ducts receive urine from the collecting tubules, and then drain into the lumen of the calyces. The epithelium lining the kidney cortex is comprised of ciliated simple cuboidal epithelium, whereas the calyx is lined with simple squamous epithelium.

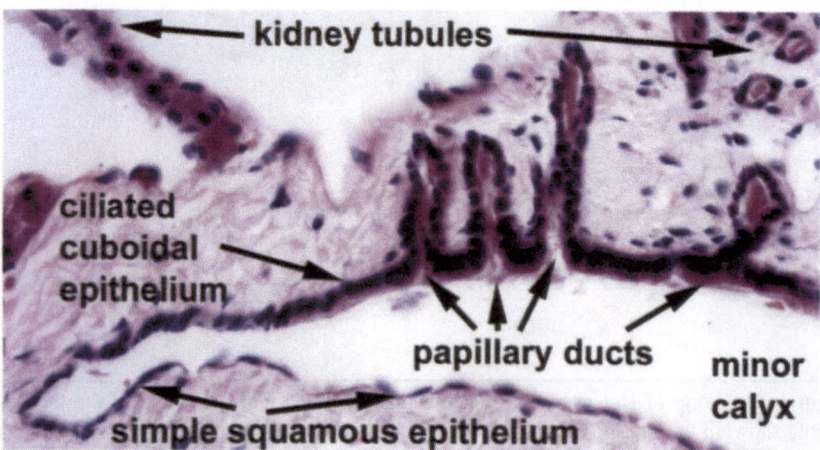

II URETER

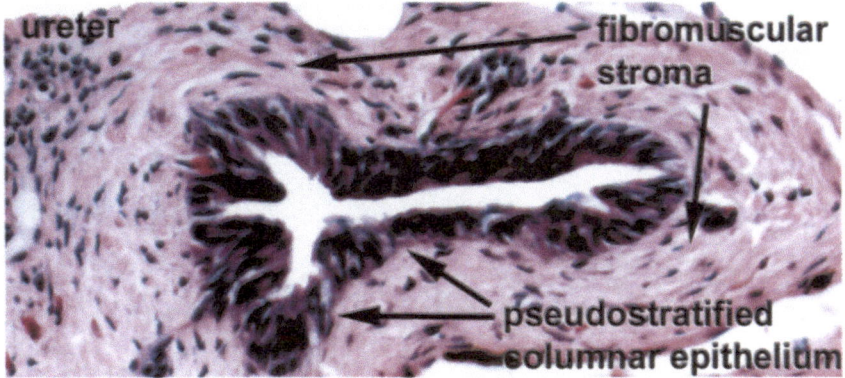

Figure 17. The proximal portion of the ureter. The ureter is a tubular structure that transports urine from the kidney pelvis to the urinary bladder. The lumen is lined with a pseudostratified columnar epithelium. A fibromuscular stroma, comprised of smooth muscle cells and dense connective tissue, surrounds the epithelium.

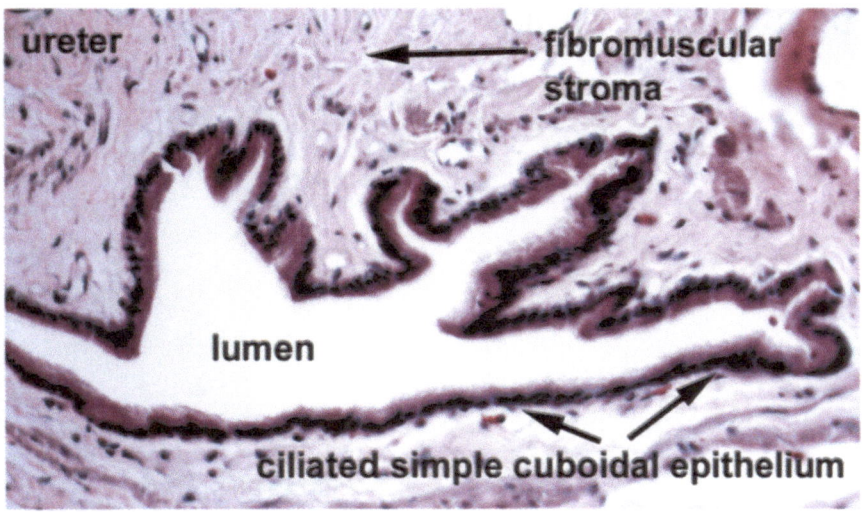

Figure 18. The distal portion of the ureter. The distal ureter is lined with a ciliated simple columnar epithelium, which exhibits numerous infoldings. A fibromuscular stroma, comprised of smooth muscle cells and dense connective tissue, comprises most of the outer region of the ureter.

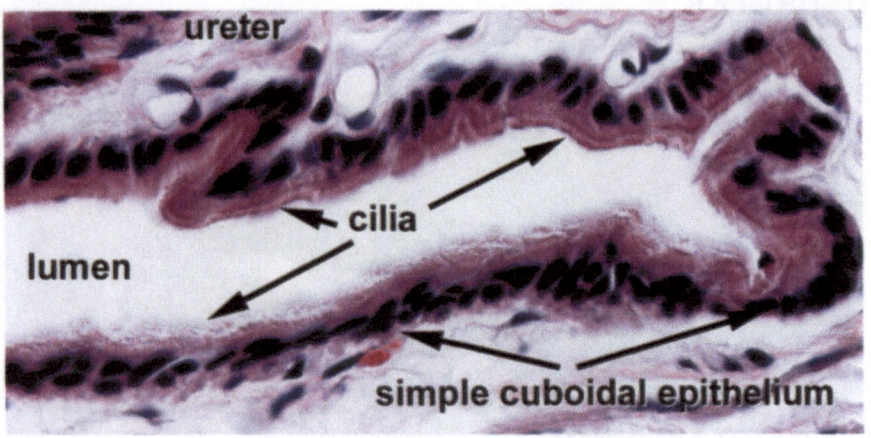

Figure 19. High magnification image of the ureter. The ureter is lined with a ciliated simple columnar epithelium. A thin layer of loose connective tissue surrounds the epithelium.

III URINARY BLADDER

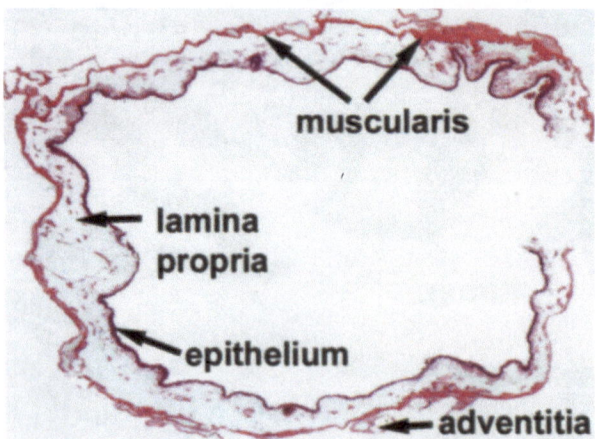

Figure 20. Low magnification image of the urinary bladder. The lumen of the bladder is lined with transitional epithelium. Underlying the epithelium is a loose connective tissue called the lamina propria. The next layer is a layer of smooth muscle cells called the muscularis. The outer layer is either an adventitia or serosa.

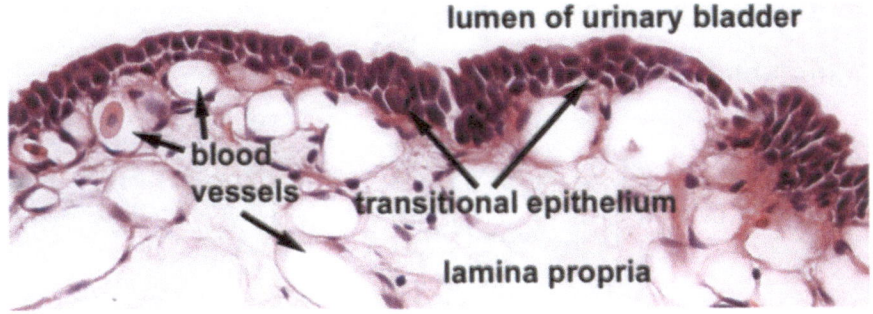

Figure 21. Epithelium and lamina propria of the urinary bladder. The layer of transitional epithelium overlies the lamina propria, which is composed of loose connective tissue. Note the presence of many blood vessels just below the epithelium basement membrane.

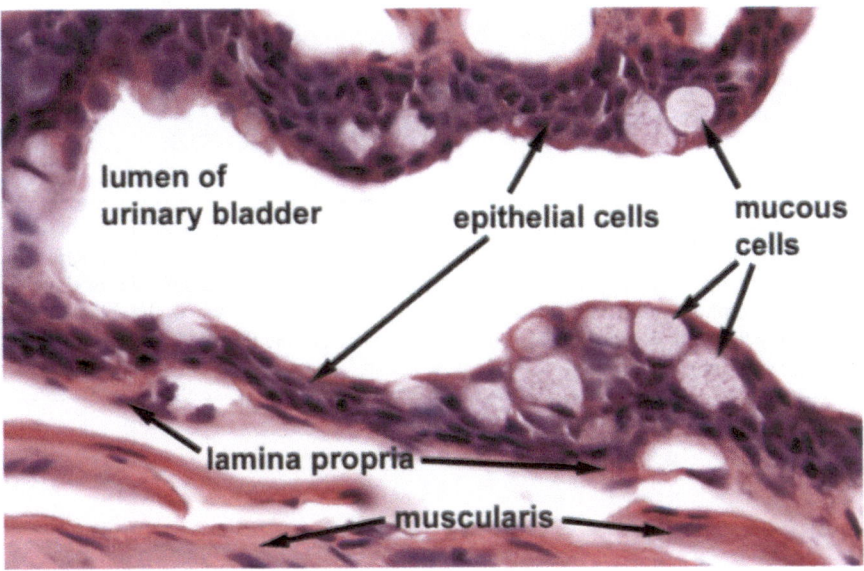

Figure 22. High magnification image of urinary bladder epithelium. The epithelium lining the lumen of the urinary bladder is transitional epithelium. Note the presence of two distinct cell types of the epithelium. The epithelial cells comprise most of the mucous lining and are stratified. Some epithelial cells appear cuboidal in shape, while other appear to be squamous. Interspersed among the epithelial cells are mucus-containing cells that have the same appearance as the flask cells of the collecting tubules of the kidney (figures 9-12, this chapter).

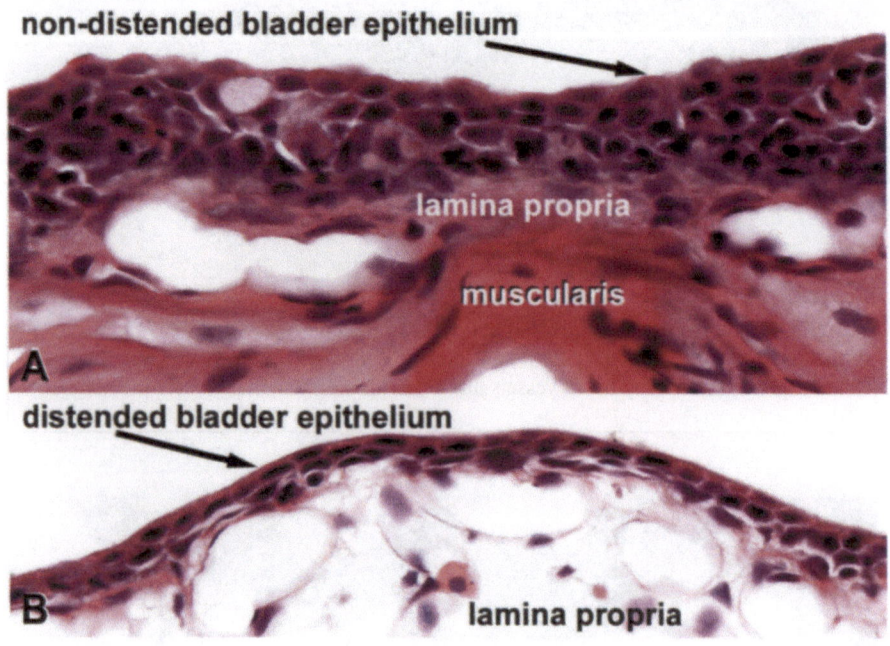

Figure 23. Transitional morphology of the urinary bladder epithelium. **A.** In the relaxed or non-distended state, when the bladder is relatively empty, the transitional epithelium appears relatively thick, with most of the cells having a cuboidal or dome-shaped morphology. **B.** In the stretched or distended state, when the bladder is relatively full of urine, the transitional epithelium is relatively thin, and the cells assume a more squamous morphology.

Chapter 7

Endocrine Organs

The organs of the endocrine system produce hormones that are released into the circulation and reach distant target tissues to exert their effects. The pituitary gland controls or influences the release of many hormones made by various endocrine organs.

I PITUITARY GLAND

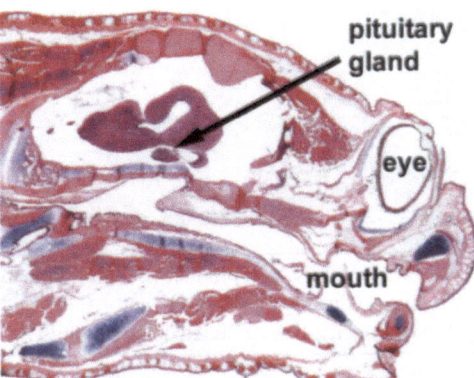

Figure 1. Low magnification of a sagittal section of a frog head, containing the pituitary gland. The pituitary gland (hypophysis) is attached to the floor of the diencephalon of the brain by an infundibular stalk. It receives hormonal signals from nuclei of the hypothalamus, which control the secretion of hormones from the pituitary gland.

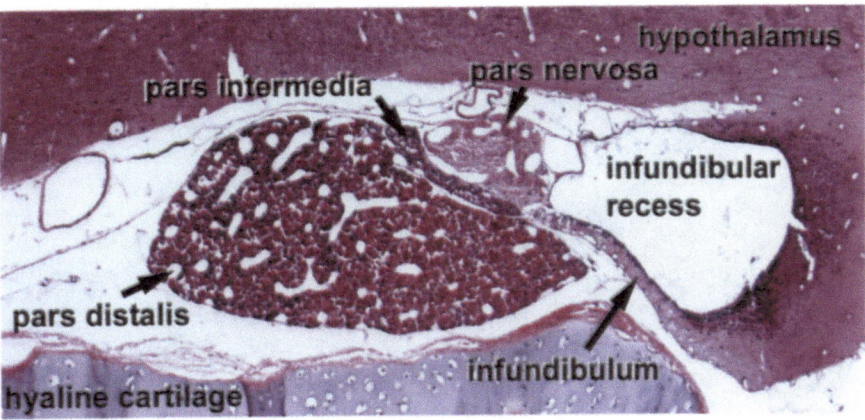

Figure 2. Pituitary gland. The pituitary gland is comprised of the adenophyophysis, which contains the pars distalis and pars intermedia, and the neurohypophysis, which contains the pars nervosa and median eminence. The pituitary gland is attached to the hypothalamus by the infundibular stalk.

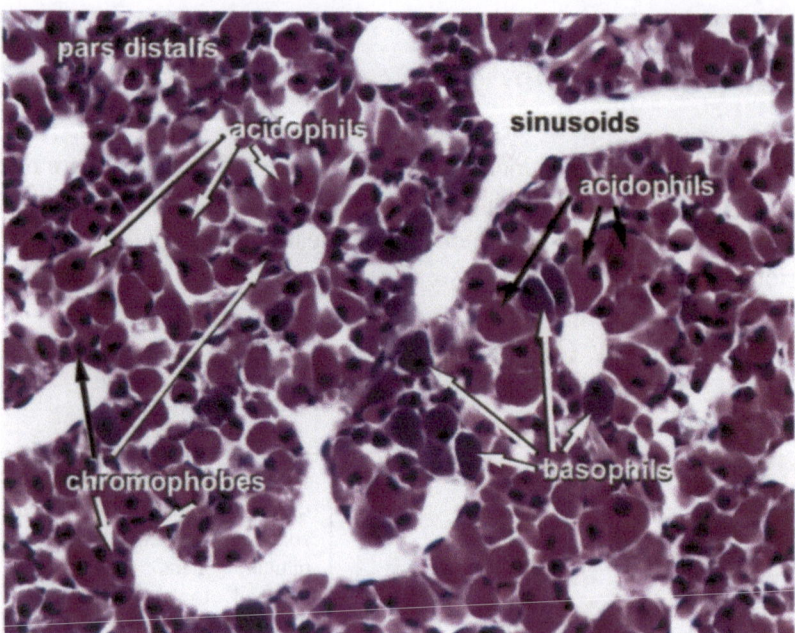

Figure 3. Pars distalis of adenohypophysis. The acidophils, which produce growth hormone and prolactin, are large eosinophilic cells. Basophils produce FSH, LH, TSF, and ACTH, and are basophilic. The chromophobes are thought to be transiently degranulated chromophils.

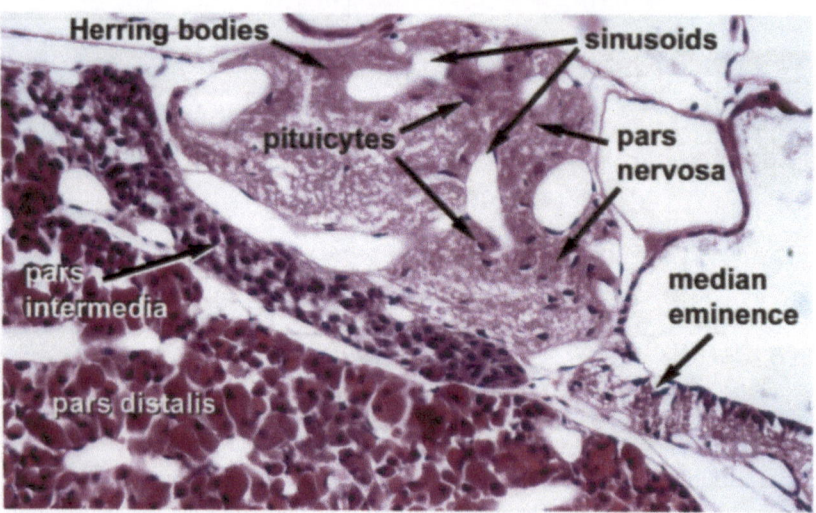

Figure 4. Three major divisions of the hypophysis. The pars intermedia appears as a sliver of cells between the pars distalis and pars nervosa. The median eminence is the region distal to the pars nervosa, which contains the primary capillary plexus where releasing hormones are secreted into the circulation to be transported to the adenohypophysis.

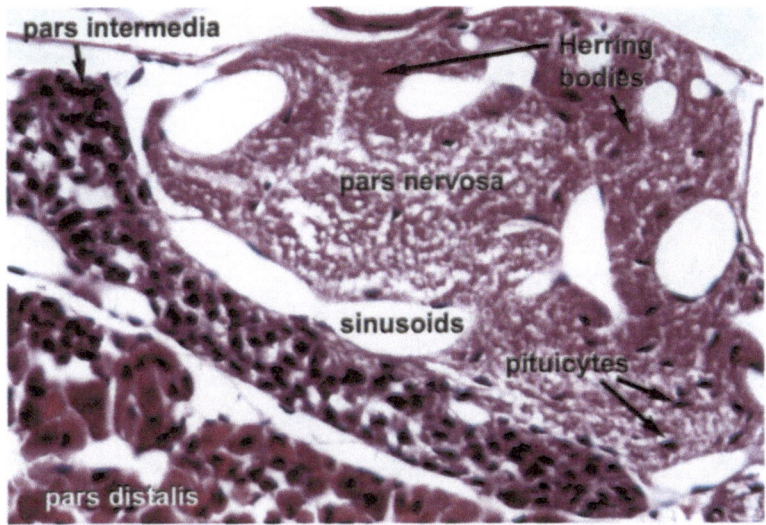

Figure 5. Pars nervosa and intermedia. Nuclei in the pars nervosa belong to pituicytes, which are astrocyte-like glial cells. Sinusoids in the pars nervosa receive hormonal secretions that are stored in the axonal swellings of the hypothalamic neurons. Cells similar to the cells of the adenohypophysis are present in the pars intermedia.

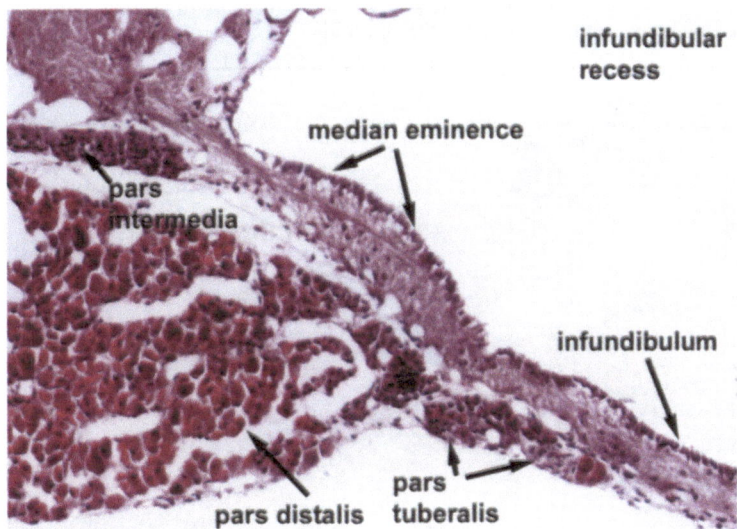

Figure 6. Median eminence. Axons of secretory neurons travel through the infundibulum to terminate at the primary capillary plexus in the median eminence where they release releasing hormones into the circulation. The releasing hormones travel through the hypophyseal portal system to the secondary plexus to stimulate release of hormones from the adenohypophysis. Axons of other neurons continue to the pars nervosa where hormones are released into the sinusoids.

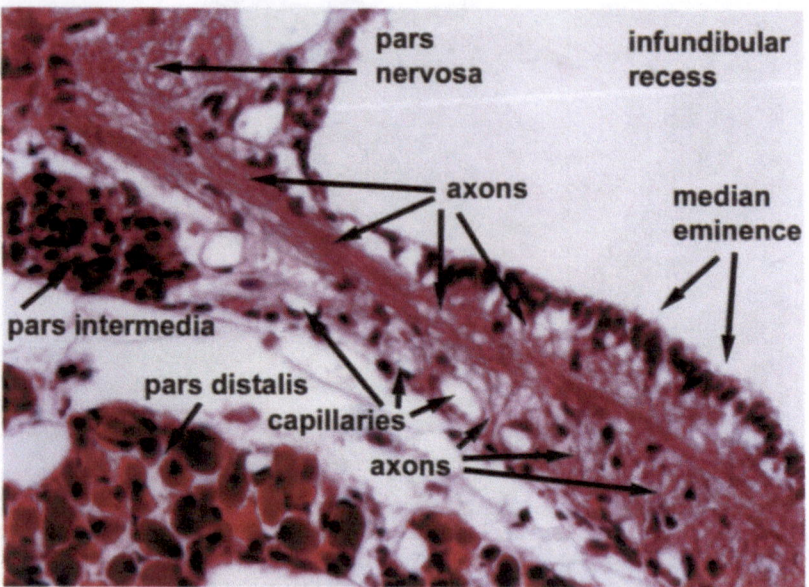

Figure 7. Median eminence. Axons of secretory neurons terminate at the capillaries of the primary plexus in the median eminence. Axons of other secretory neurons continue to the pars nervosa where hormones are stored in Herring bodies until released into the sinusoids.

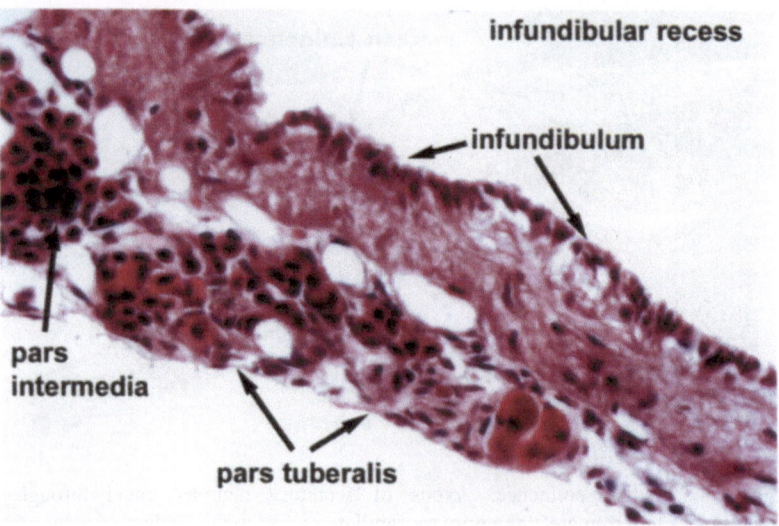

Figure 8. Pars tuberalis. The pars tuberalis is part of the adenohyphophysis and is continuous with the pars distalis. It is closely associated with the infundibulum, and is comprised of cells similar to those in the pars distalis.

II PINEAL GLAND

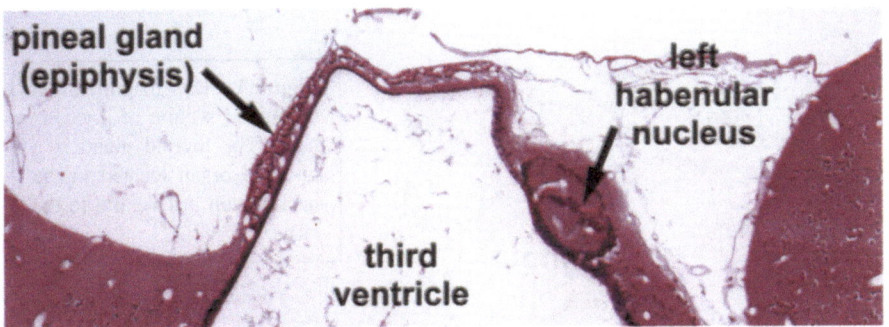

Figure 9. Low magnification image of the roof of the diencephalon containing the pineal gland. The pineal gland (epiphysis) is attached to the brain in the region of the left and right habenular nuclei.

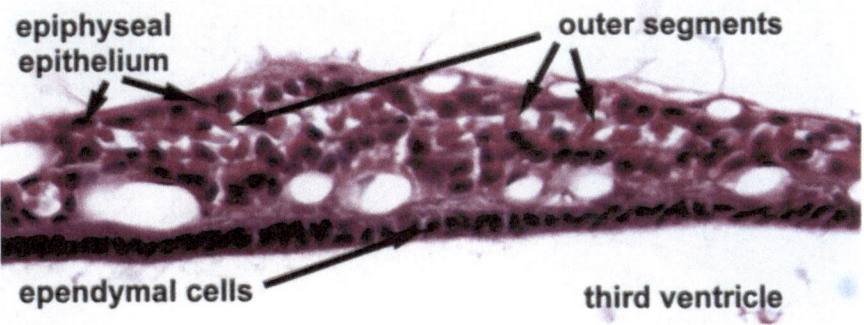

Figure 10. High magnification of the pineal gland. Ependymal cells line the third ventricle. The principal cells of the pineal gland are pinealocytes (also called epiphyseal epithelium), which produce the hormone melatonin. In *Xenopus*, the pinealocytes have an appearance similar to retinal photoreceptors with outer segments that protrude into the lumens of the gland. Interstitial cells are glial-like cells and are identified by their elongate or irregular-shaped nuclei.

III THYROID GLAND

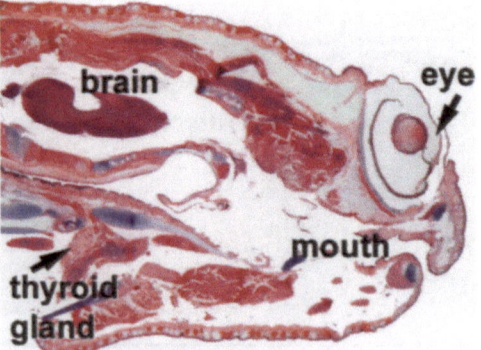

Figure 11. Low magnification of sagittal section of *Xenopus* head. The thyroid gland is a bi-lobed organ located in the neck region just ventral to the larynx.

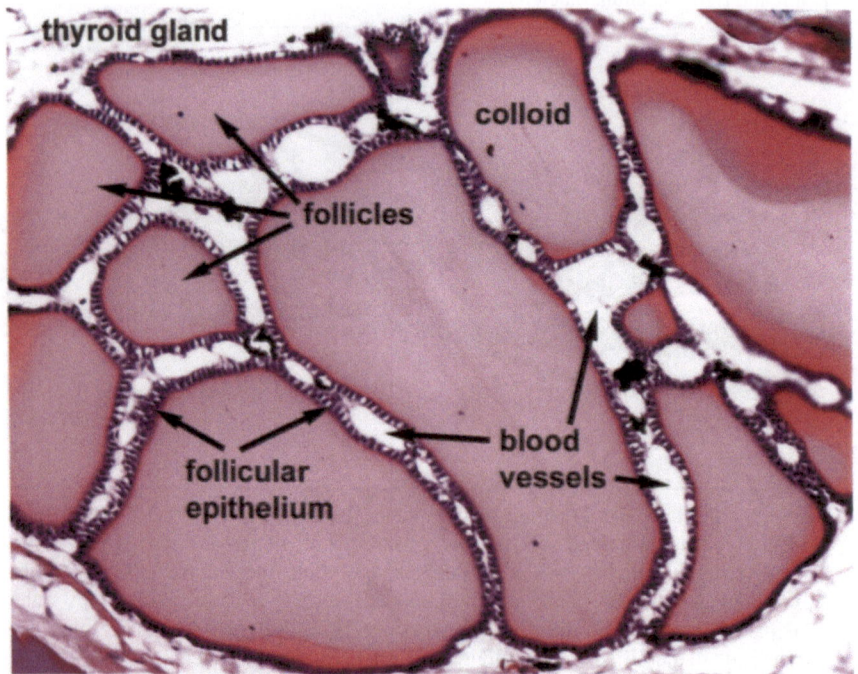

Figure 12. Intermediate magnification image of the thyroid gland. The thyroid gland is composed of spherical colloid-filled follicles that are lined by follicular cells. The follicular cells produce thyroid hormones by the production of thyroglobulin, which is stored in the extracellular colloid. At the basal side of the follicular cells is an extensive capillary plexus.

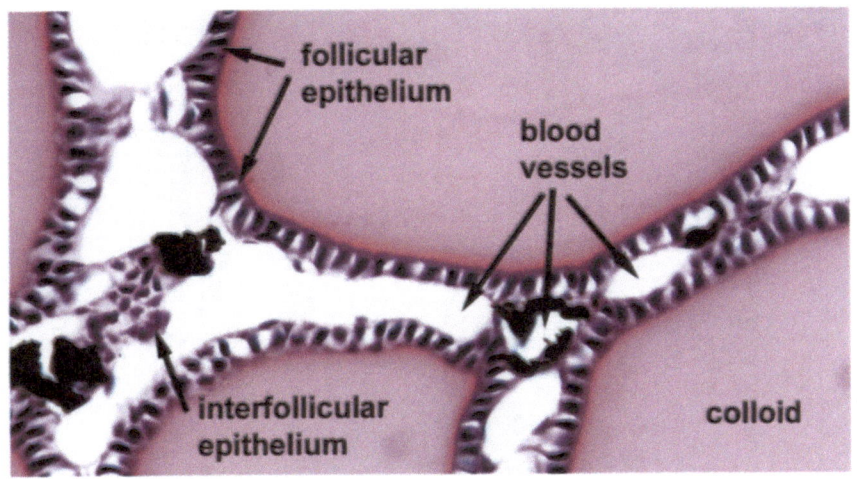

Figure 13. High magnification image of the thyroid gland. The follicular cells that line the follicles are cuboidal or columnar depending on their level of activity which is controlled by the release of thyroid stimulating hormone from the pars distalis. There is an extensive capillary network at the basal surface of the follicular cells. There are some groups of cells in clusters between the follicles, called interfollicular cells. These cells are also referred to as "C" cells because they secreted the hormone calcitonin into the circulation.

IV ADRENAL (INTERRENAL) GLAND

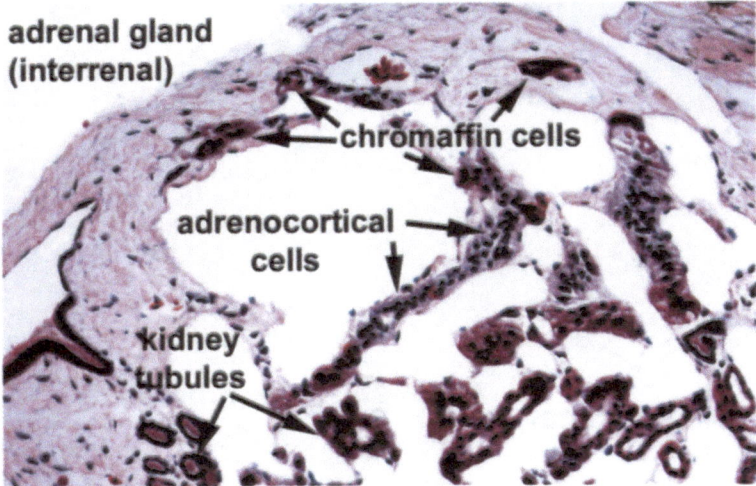

Figure 14. Low magnification image of the adrenal gland. The adrenal gland of *Xenopus laevis* is also called the interrenal gland because of its position in the medial regions of the kidneys where the kidneys are apposed to each other. The adrenal gland appears as clusters of chromaffin cells and adrenocortical cells just within the renal capsule.

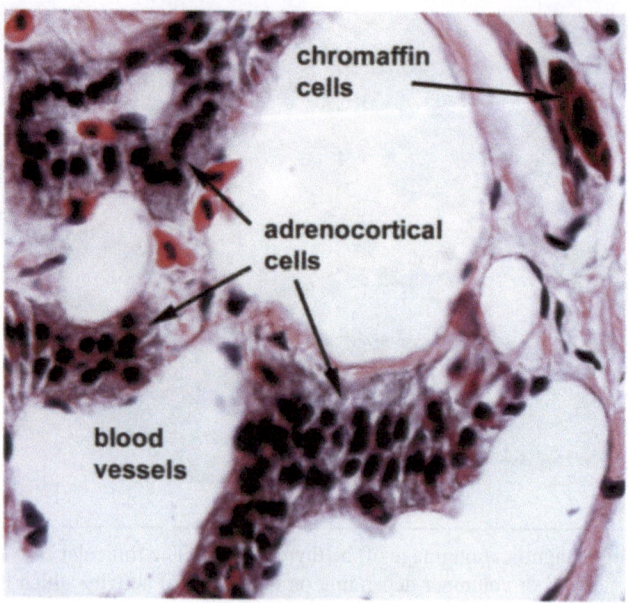

Figure 15. High magnification image of the adrenal gland. The steroid-secreting adrenocortical cells comprise the major component of the adrenal gland and are arranged in cords. There are many pale vacuoles in the cytoplasm of the adrenocortical cells that represent the numerous cholesterol-containing lipid droplets. The catecholamine-secreting chromaffin cells are arranged singly or in small clusters.

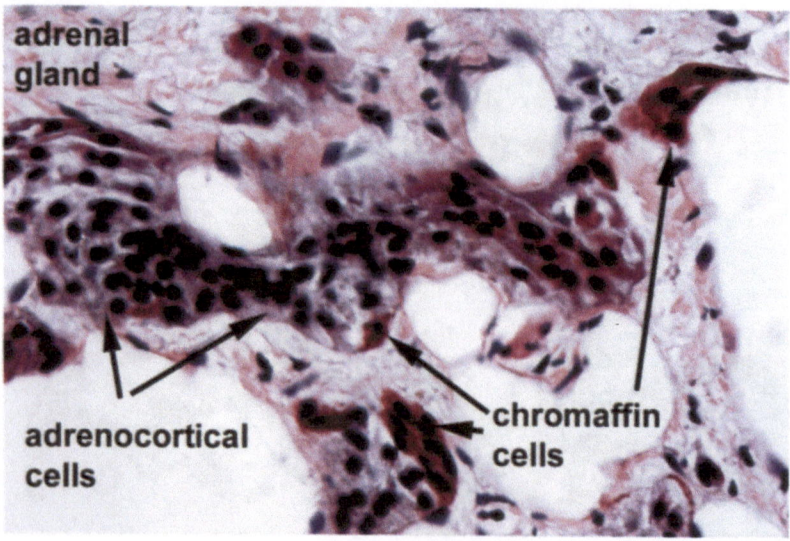

Figure 16. High magnification image of the adrenal gland. The reddish-brown staining chromaffin cells are arranged singly or in small clusters and are often closely associated with the light purple-staining adrenocortical cells.

Chapter 8

Reproductive Organs

The reproductive organs function to produce and transport germinal cells (oocytes and sperm) to a site where fertilization can occur. The male reproductive organs are the testes, and the female reproductive organs are the ovaries. The oviduct serves to deliver the oocytes from the ovaries to the external environment, where they will be fertilized by the sperm cells.

I OVARY

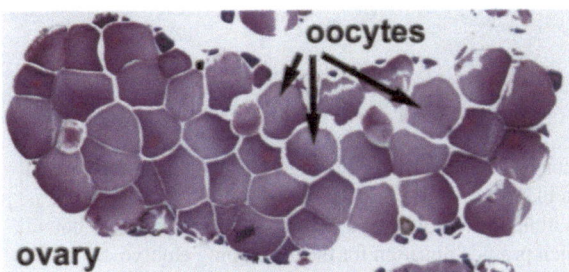

Figure 1. Low magnification image of the ovary. The ovary is comprised primarily of many oocytes clustered together.

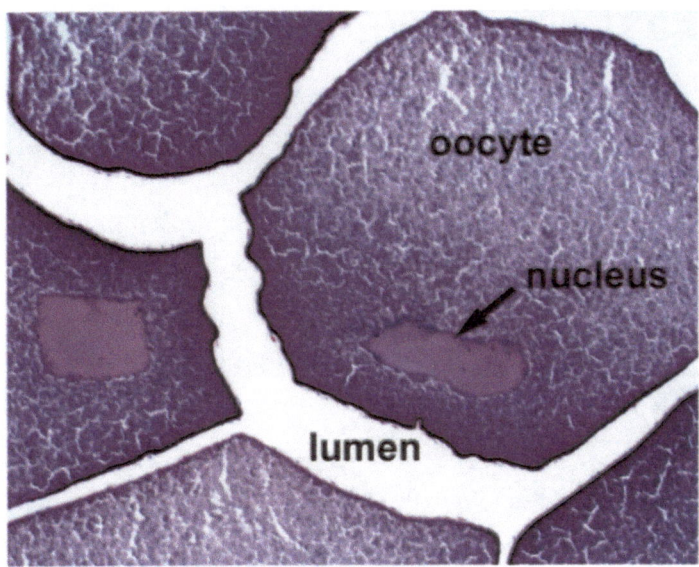

Figure 2. Oocytes comprise most of the ovary. Mature oocytes have a diameter of about 1.5 mm. A large nucleus can be observed in most oocytes, and a layer of pigment granules closely underlie the plasma membrane.

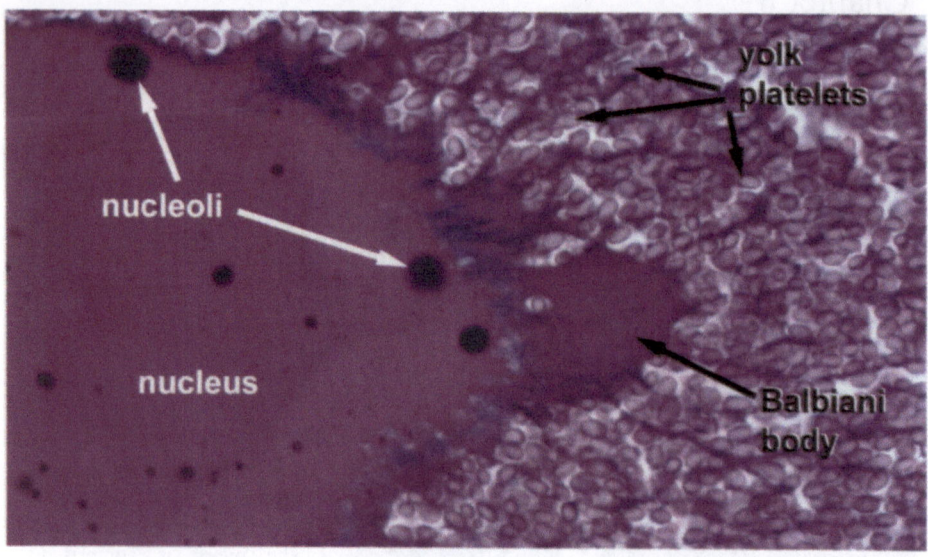

Figure 3. Oocyte nucleus. The nuclei of oocytes have many nucleoli that tend to be located near the nuclear envelope. A Balbiani body, which is a large mass of mitochondria, is located close to the nucleus. Within the cytoplasm are many small membrane-bound structure called yolk platelets, which provide nutrition for the developing embryo.

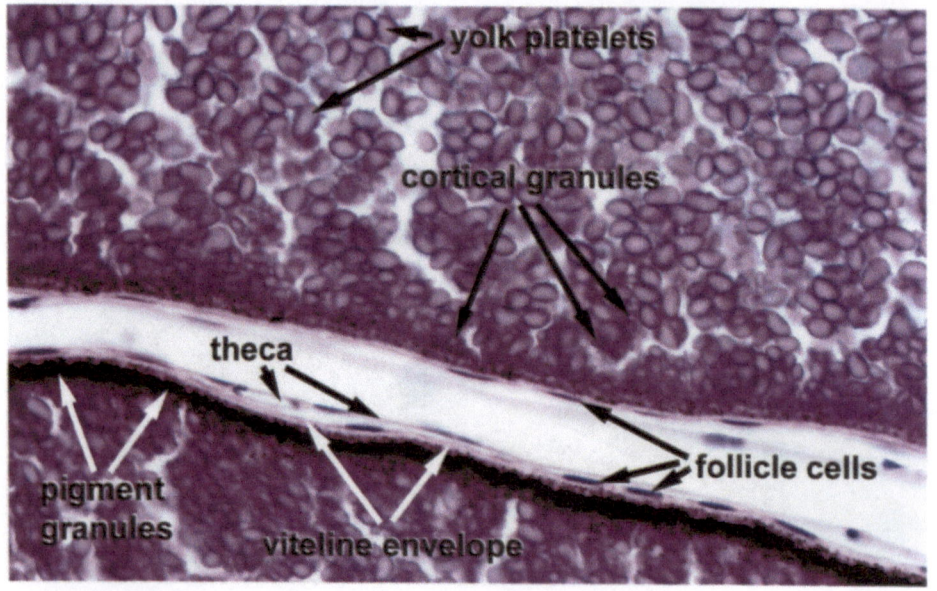

Figure 4. Oocyte cytoplasm and membrane. Cortical granules are located near the plasma membrane, and a layer of pigment granules closely underlie the membrane. Flattened follicular cells are on the extracellular side of the oocyte membrane. A delicate theca layer overlies the follicular cells.

II OVIDUCT

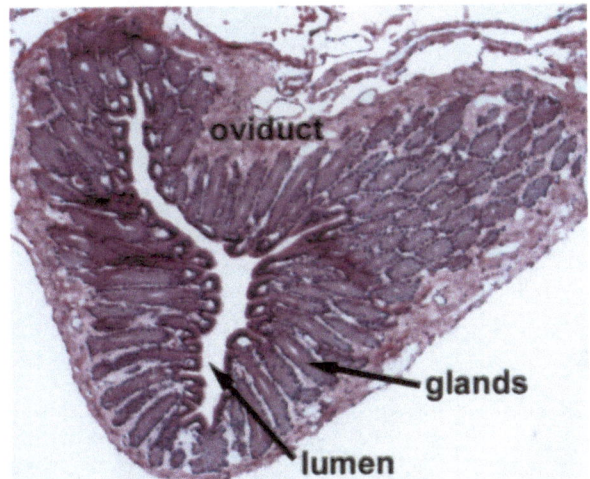

Figure 5. Low magnification image of a cross section of the oviduct. The lumen of the oviduct is lined by many glands that secrete into the lumen. A fibrous stroma supports the basal portion of the glands.

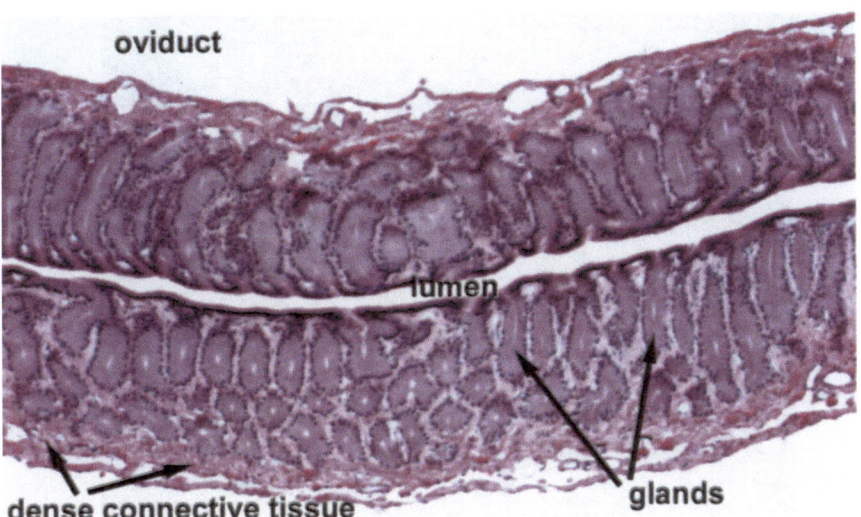

Figure 6. Longitudinal section of the oviduct. The oviduct is surrounded by a layer of dense connective tissue. Long glands empty their secretory products into the lumen as oocytes travel along the lumen from the ovary. The glandular secretions form the layers of jelly coats on the oocytes as they are transported to the cloaca.

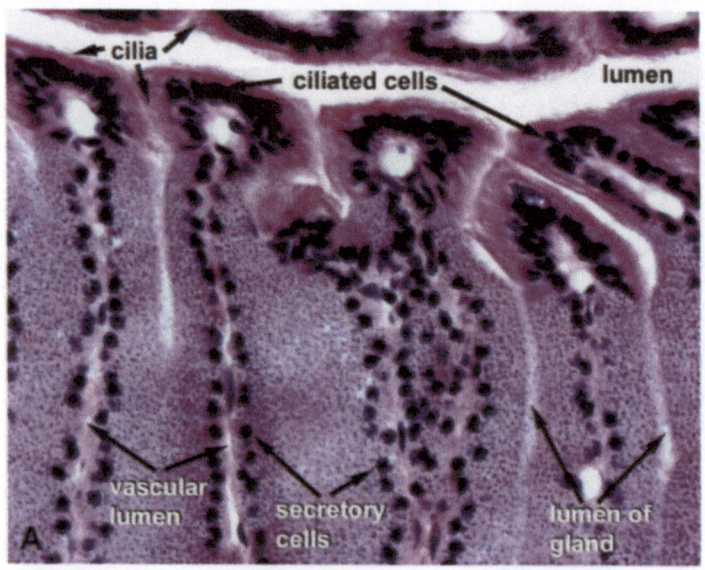

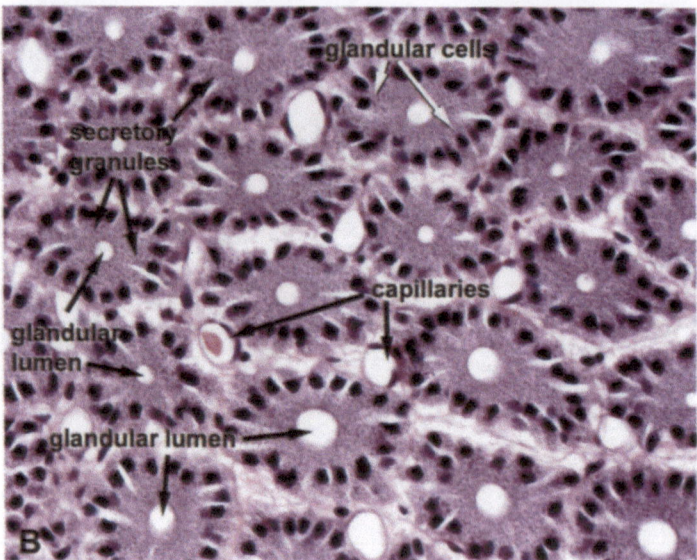

Figure 7. High magnification images of oviduct glands. **A.** Longitudinal orientation of the oviduct glands. The glandular cell nuclei are located at the basal side of the cells and secrete their contents into the glandular lumen that is confluent with the lumen of the oviduct. The epithelial cells lining the oviduct lumen are ciliated, and aid in the transport of oocytes from the ovaries to the cloaca. Note the vascular supply between the glands. **B.** Horizontal orientation of the oviduct glands. The glandular lumen is located centrally, and the glands are surrounded by loose connective tissue with many capillaries. Note the dark-staining secretory granules that contain mucin-like glycoproteins that are secreted to make the layers of jelly coats on the oocytes as they are transported through the lumen.

III TESTES

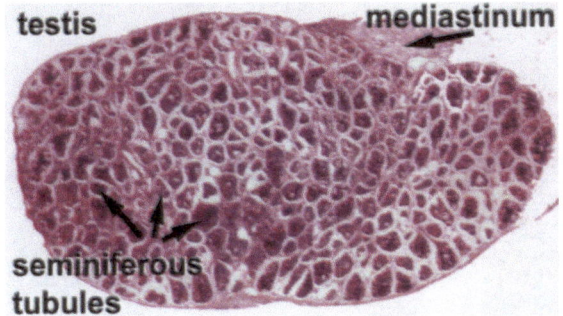

Figure 8. Low magnification image of the testes. The testes are surrounded by a dense connective tissue capsule. The mediastinum receives sperma-tids that are produced in seminiferous tubules, which comprise most of the testes.

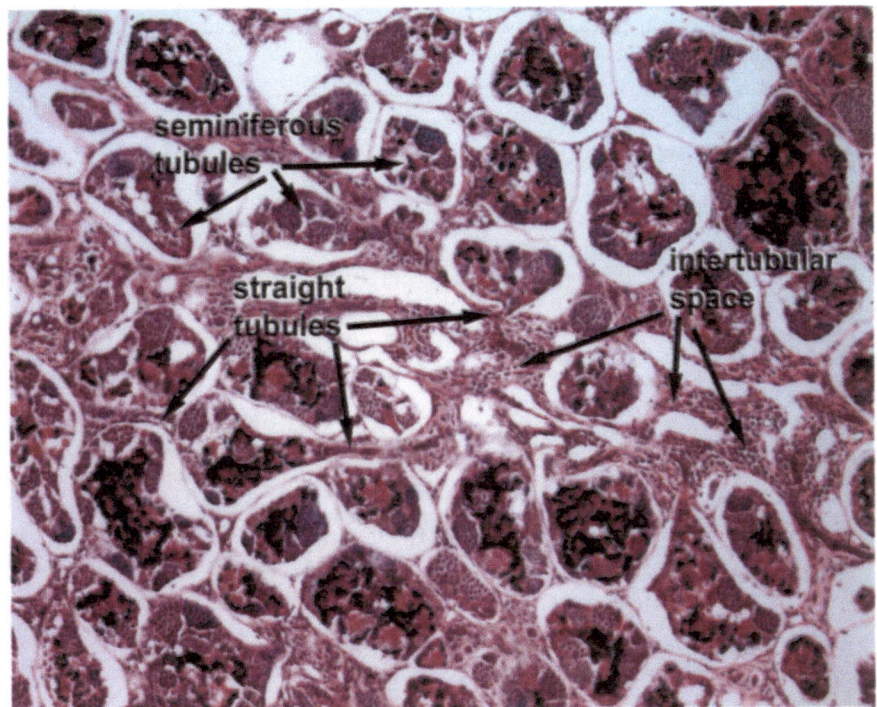

Figure 9. Seminiferous tubules of the testes. The seminiferous tubules are surrounded by loose connective tissue and this area is called the intertubular space. In this speciment the space between each seminiferous tubule and the connective tissue of the intertubular space is an artifact of the paraffin embedding process. The seminiferous tubules are confluent with the straight tubules (also called tubuli recti) through which the spermatids travel to the rete testes of the mediastinum and from there will exit the testes.

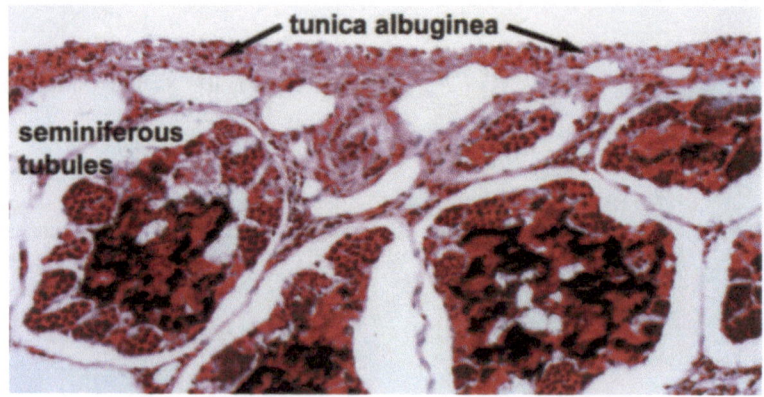

Figure 10. Tunica albuginea. A layer of dense connective tissue, called the tunica albuginea, encapsules the testes. Directly beneath the tunica albuginea are the seminiferous tubules.

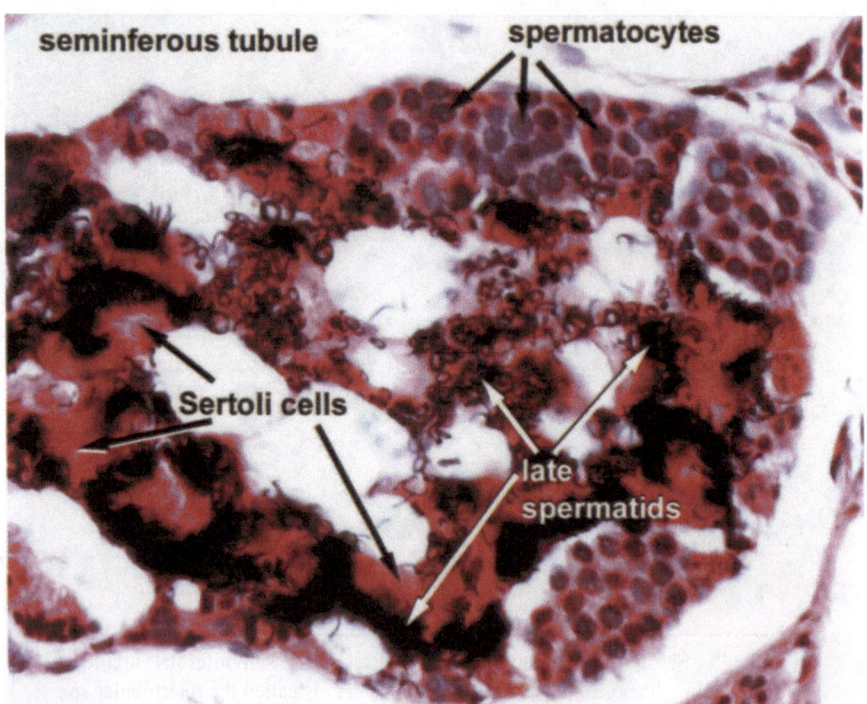

Figure 11. Seminiferous tubules of the testes. Spermatocytes are arranged in clusters near the basement membrane. Sertoli cells are large, eosinophilic cells and have many late spermatids attached to their apical surface. After some maturation of the spermatids, they detach from the Sertoli cells, and pass through the seminiferous tubules into the straight tubules.

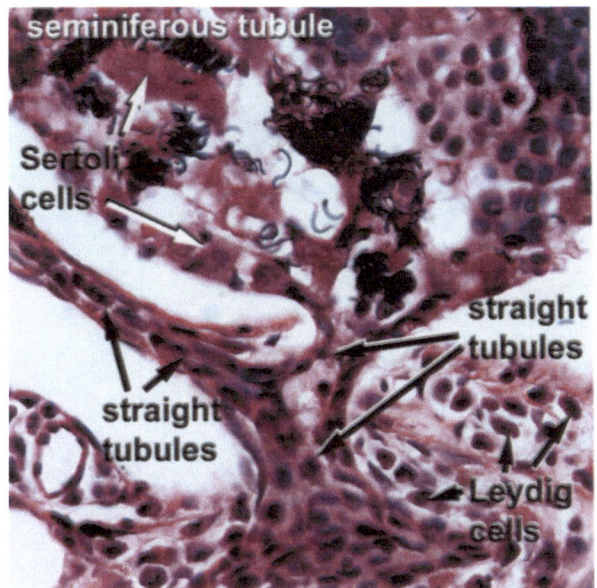

Figure 12. Seminiferous tubules and straight tubules. The seminiferous tubules are in continuity with the straight tubules. The sraight tubules receive the sperma-tids from the seminiferous tubules and transport them to the mediastinum. Note the branching of the straight tubules as they join to form a larger tubule. Leydig cells are in the interstitial con-nective tissue.

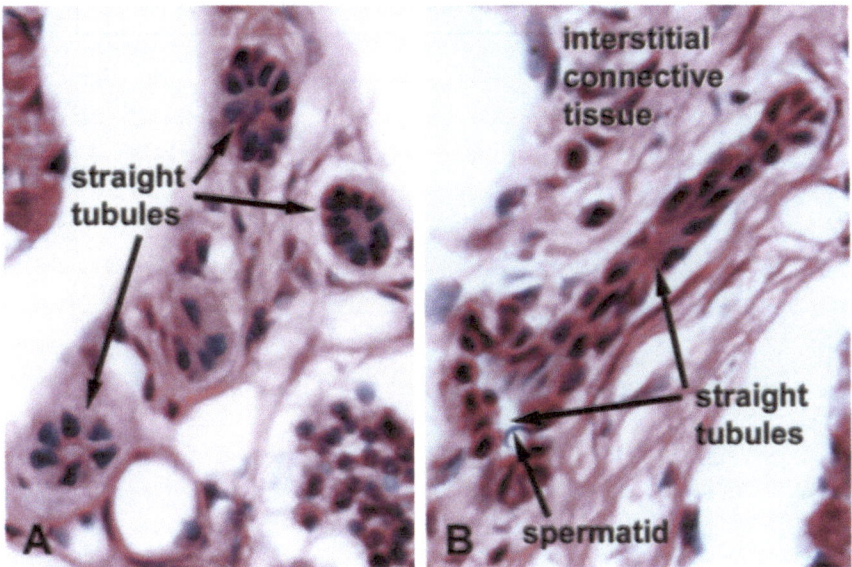

Figure 13. Straight tubules. Straight tubules (tubuli recti) are located within the interstitial connective tissue. **A.** Straight tubules in cross-section. The tubules are lined by a cuboidal epithelium and have a small lumen. B. Tangential section of a straight tubule at a branching point. Note the presence of a spermatid in the lumen.

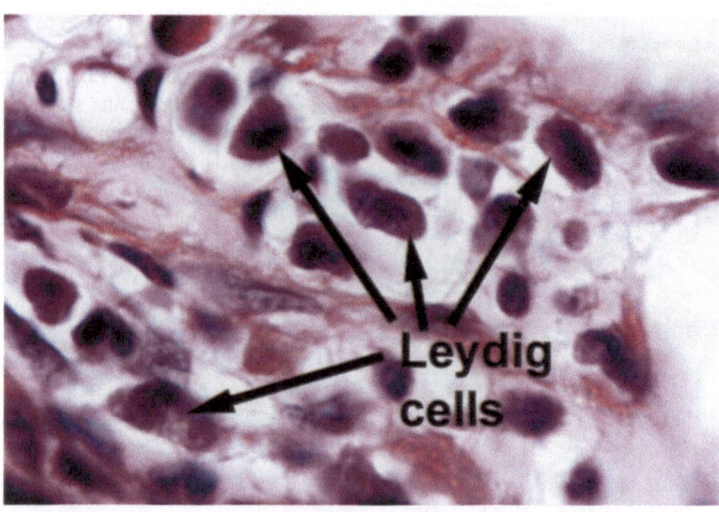

Figure 14. Leydig cells. The interstitial cells of Leydig are located the interstitial space between the seminiferous tubules. Leydig cells produce male androgens such as testosterone. They can be identified by their irregularly shaped nuclei and vacuolated cytoplasm. The vacuoles are the result of extraction of lipid droplets in the cytoplasm.

Chapter 9

Integument

The integumentary system consists of the skin (integument) and its associated structures. The skin is composed of two major portions; the epidermis and the dermis. In *Xenopus*, the epidermis is a keratinized stratified squamous epithelium. The dermis is composed of loose and dense connective tissue. Associated with the skin are exocrine glands which secrete a variety of substances through ducts to the body surface, and lateral line organs, which are sensitive to fluid movement.

I EPIDERMIS

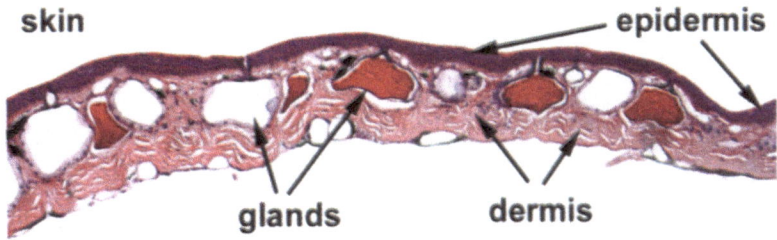

Figure 1. Low magnification image of *Xenopus* skin. The skin (integument) is composed of the epidermis and dermis. Within the dermis are many glands that secrete material to the surface.

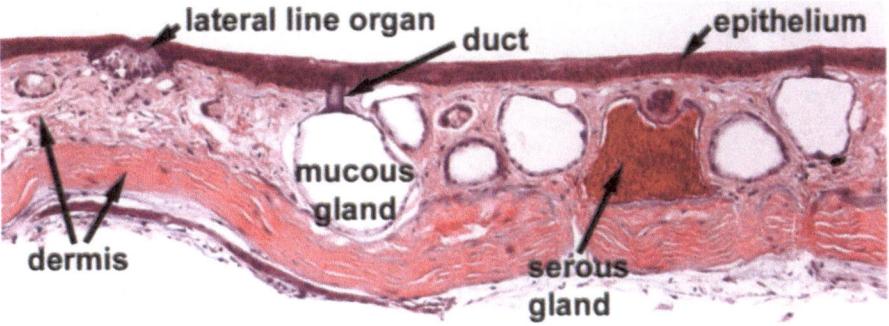

Figure 2. The integument of *Xenopus laevis*. Lateral line organs protrude from the epithelium to the surface. Mucous glands and serous glands secretes materials through excretory ducts to the surface. The dermis is composed of dense connective tissue.

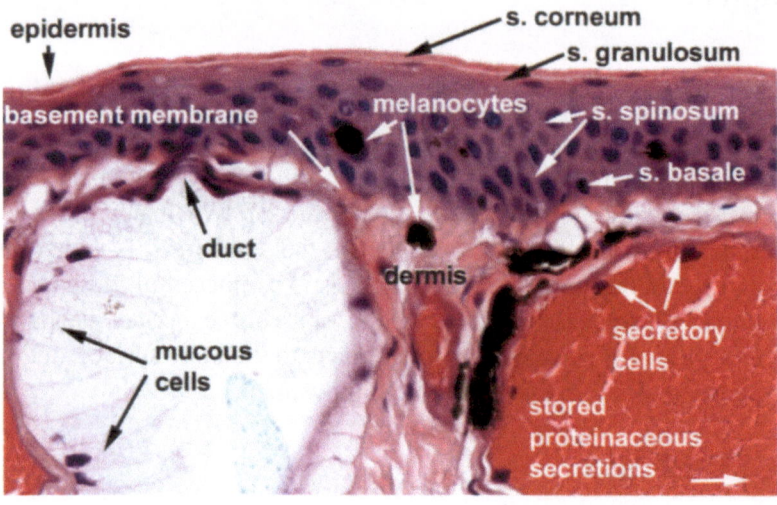

Figure 3.   The epidermis. The epithelium of the epidermis is a keratinized stratified squamous epithelium, and is composed of four distinct layers; the stratum corneum, stratum granulosum, stratum spinosum, and stratum basale. A thin stratum lucidum also appears between the corneum and granulosum. Melanocytes are present throughout the epithelium.

II DERMIS

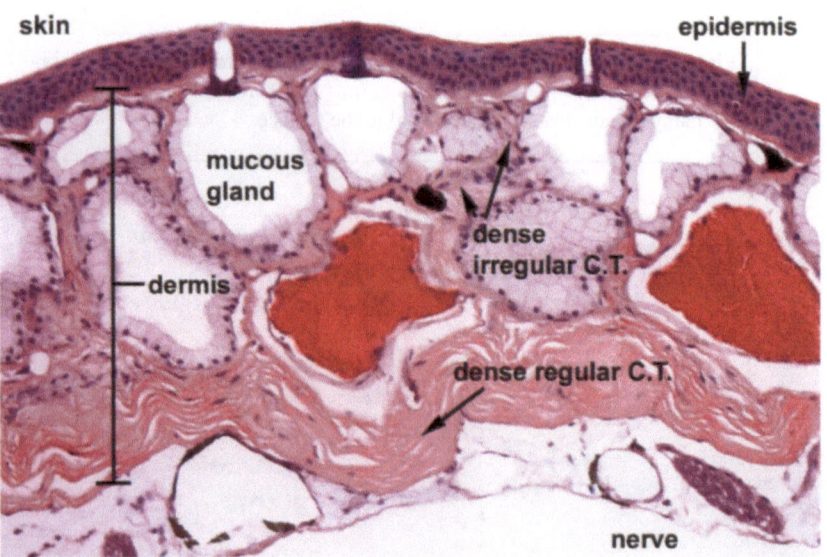

Figure 4.   The dermis. Mucous and serous glands are prominent in the dermis, and expel their secretory contents to the surface of the epidermis through excretory ducts. The dense connective tissue (CT) of the dermis is dense irregular CT, which is close to the epidermis, and the CT of the more distal region of the dermis is composed of dense regular CT.

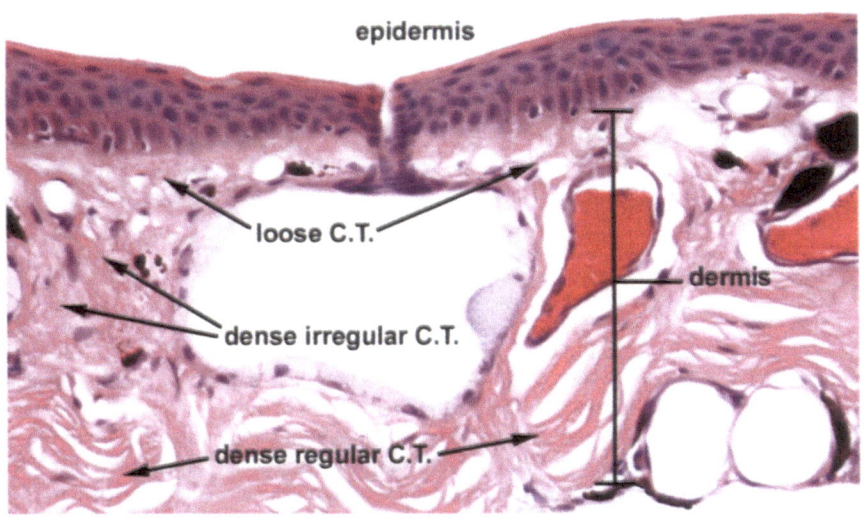

Figure 5. The dermis. Directly underlying the epidermis is the loose connective tissue (CT) of the dermis, which contains many capillaries. Deep to the loose CT is the dense irregular CT of the dermis, and deep to that is the dense regular CT of the dermis.

III GLANDS

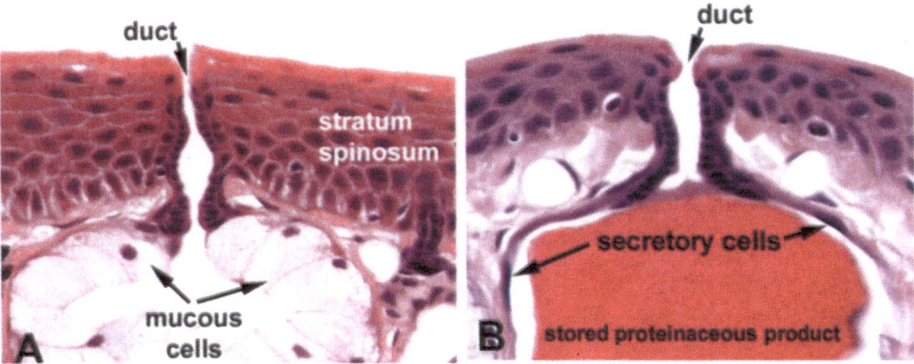

Figure 6. Exocrine glands of the skin. **A.** Mucous glands are prominent in the integument and secreted mucus through an excretory duct composed of cuboidal epithelium. The mucous cells have a highly vacuolated cytoplasm and a round nucleus. **B.** The protein-secreting glands of the skin (also called granular cells) secrete their proteinaceous products through an excretory duct composed of cuboidal epithelium. Unlike the mucous gland cells, these granular cells constitute a syncytium (fusion of several cells).

IV   LATERAL LINE ORGAN

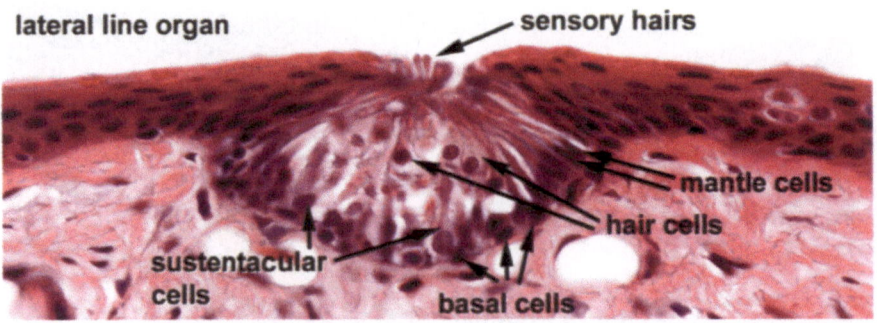

Figure 7. Lateral line organs of the skin. The lateral line organs are small, elongated structures on the head and trunk of *Xenopus*. Hair cells on the surface are called neuromasts an respond to a flow of water over the surface. Each neuromast consists of a group of cells embedded in the epidermis. The mantle cells lie peripherally and surround the centrally positioned sensory and supporting sustentacular cells. The sustentacular cells extend from the basement membrane to the outer surface. The sensory hair cells extend to the free surface, but to not reach the basement membrane. The sensory hairs of the hair cells consist of one kinocilium and 20-40 stereocilia which are enclosed by a gelatinous cupula. The basal cells are attached to the basement membrane but do not reach the free surface.

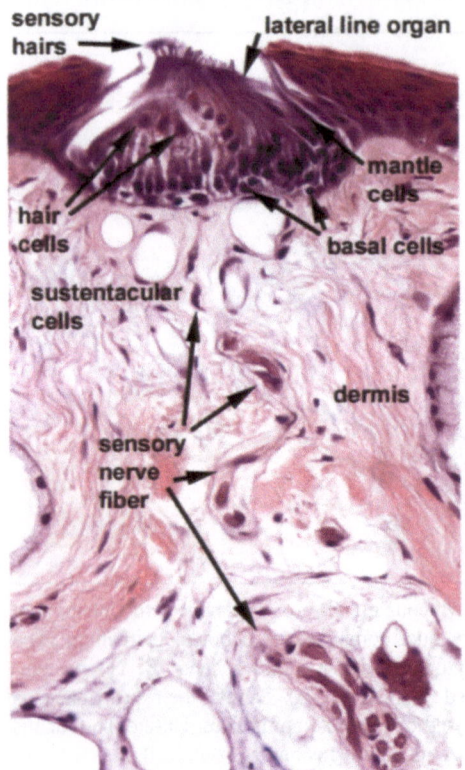

Figure 8. Innervation of lateral line organs. The lateral line organs are embedded in the epidermis, and are innervated by sensory nerve fibers that pass through the dermis. The lateral line sensory system functions to detect and localize movements of nearby obstacles, prey, predators, and con-specifics in the environment.

Chapter 10

Cranial Structures

Cranial structures illustrated in this chapter are: oral cavity including the oral mucosa, taste buds, salivary glands and teeth; epiglottis; esophagus; nasal cavity; brain; inner ear; and eye and surrounding supporting structures including Harderian glands. The cranium is composed of all four tissue types and most of the main neural structures contain or receive projections from primary sensory neurons. The skin of the cranium is discussed in Chapter 9. Primary sensory cells are located in taste buds, retina, olfactory and vomeronasal chemoreceptive epithelium, auditory and vestibular labyrinth in the inner ear, and sensory ganglia of the cranium such as the somatosensory trigeminal ganglion.

I   OVERVIEW OF CRANIAL STRUCTURES

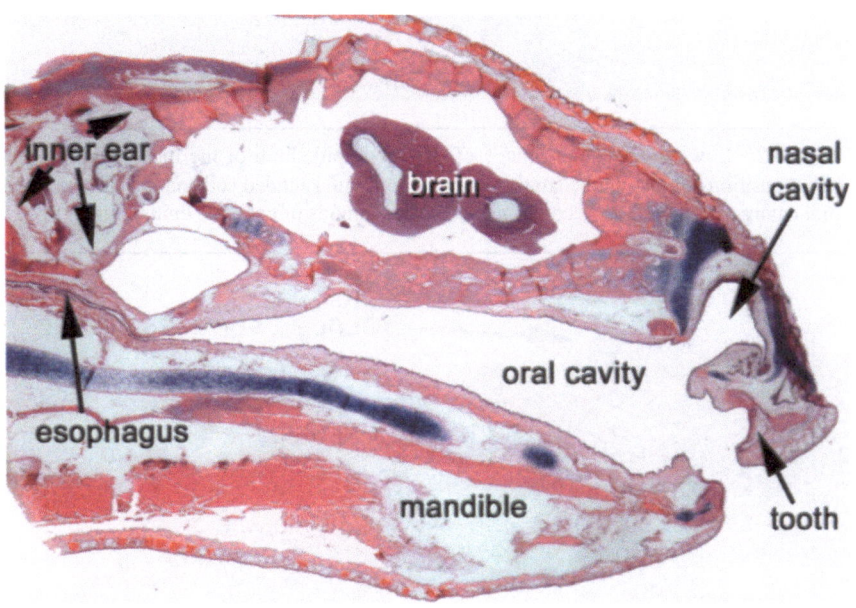

Figure 1.   Low magnification image of the *Xenopus* cranium in the parasagittal plane. The rostral side of the head is to the right.

II ORAL CAVITY

There is a unique stratified epithelium on the ventral lip region. The oral cavity is lined with stratified epithelium. Tooth structures are present underlying the epithelium on the inner rostral and lateral margins of the maxillary lip. *Xenopus* frogs do not have tongues. Taste buds are distributed over the palate and floor of the oral cavity.

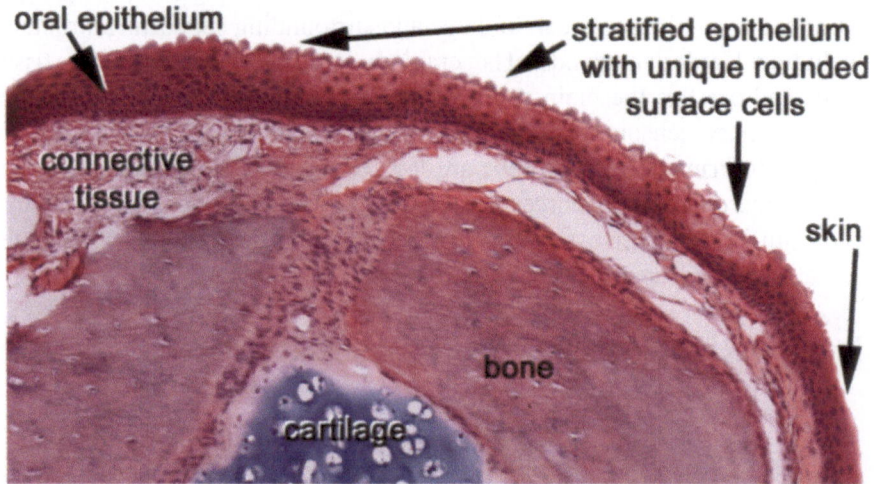

Figure 2. Low magnification image of stratified epithelium of the mandibular lip. The skin transitions into a unique stratified epithelium with rounded cells on the surface. In the oral cavity the epithelium becomes a stratified sqaumous or cuboidal epithelium.

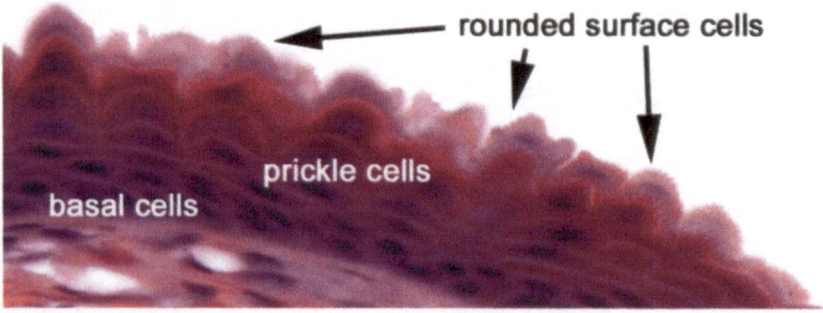

Figure 3. High magnification of transitional area of lip. This area is a stratified epithelium with typical basal and prickle cells, but there is a unique class of rounded surface cells that possess numerous extensions or blebs of membrane protruding from their surface.

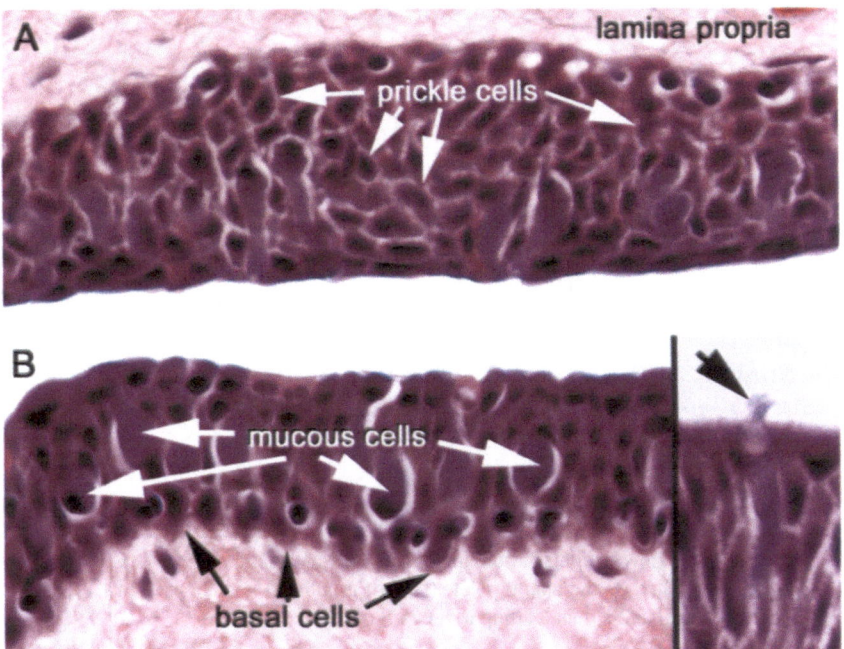

Figure 4. (ABOVE) High magnification image of mucosal lining of the oral cavity. **A.** Nonkeratinized stratified squamous epithelium as seen on the palate. **B.** Stratified cuboidal epithelium as seen on the floor of the mouth. Both epithelia contain a layer of basal cells that give rise to new epithelial cells, elongated mucous cells that project to the surface and typical prickle cell keratinocytes. The connective tissue underlying the epithelium is called the lamina propria. This layer contains mostly fibroblasts and collagen fibers. **Inset** in **B** shows mucus secreted from a mucous cell.

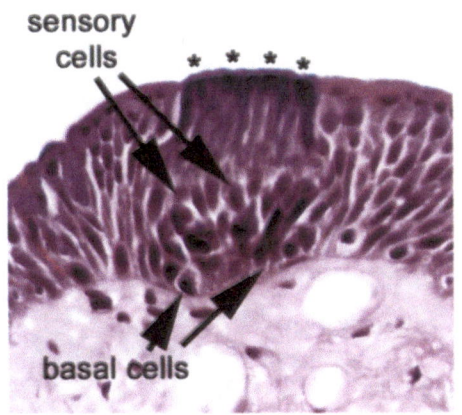

Figure 5. (LEFT) High magnification image of a taste bud. Taste buds are located on the palate and floor of the oral cavity. They consist of a round collection of basal, sensory and supporting cells. The receptive end of the sensory cells (basophilically stained dark purple) projects to the surface of the epithelium. Taste molecules stimulate the cells at their sensory ending (asterisks).

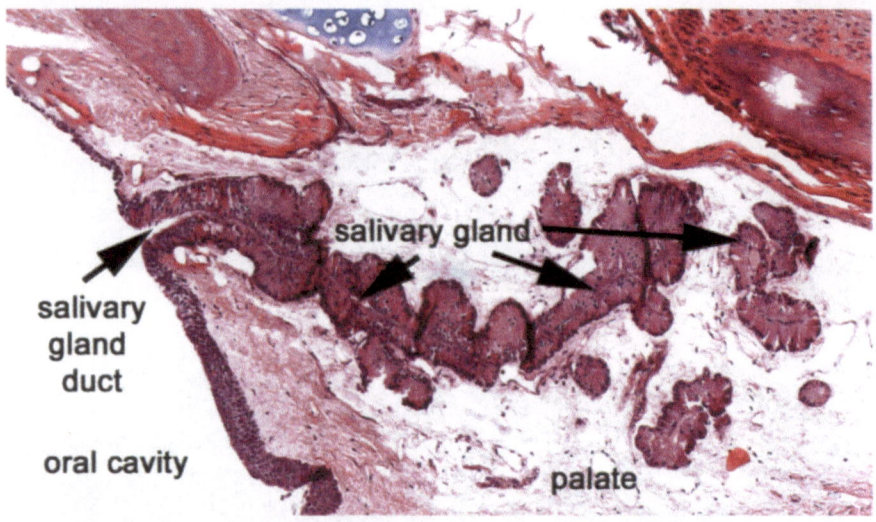

Figure 6. Low magnification image of salivary gland. A salivary gland is located on the lateral palate (roof of mouth). It is composed of mucous acini and a duct leading to the oral cavity.

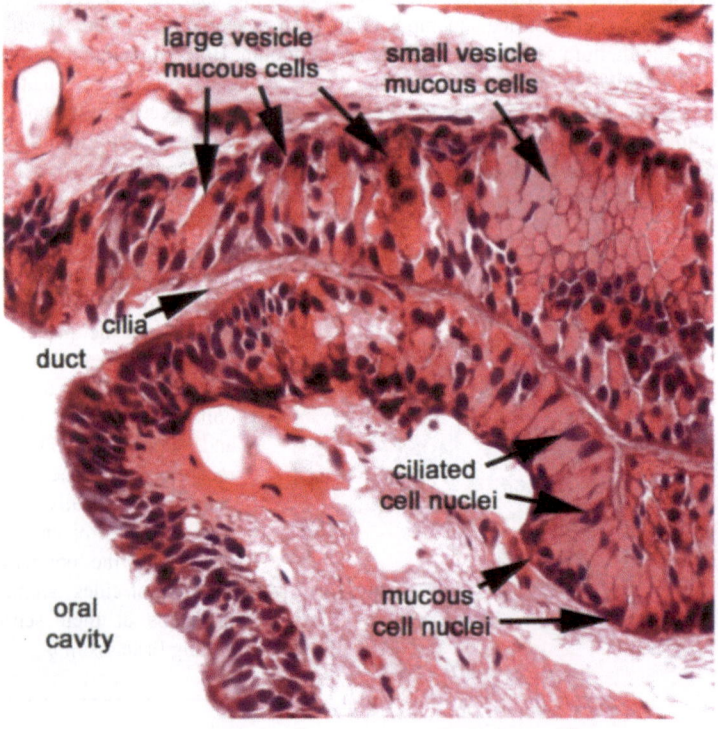

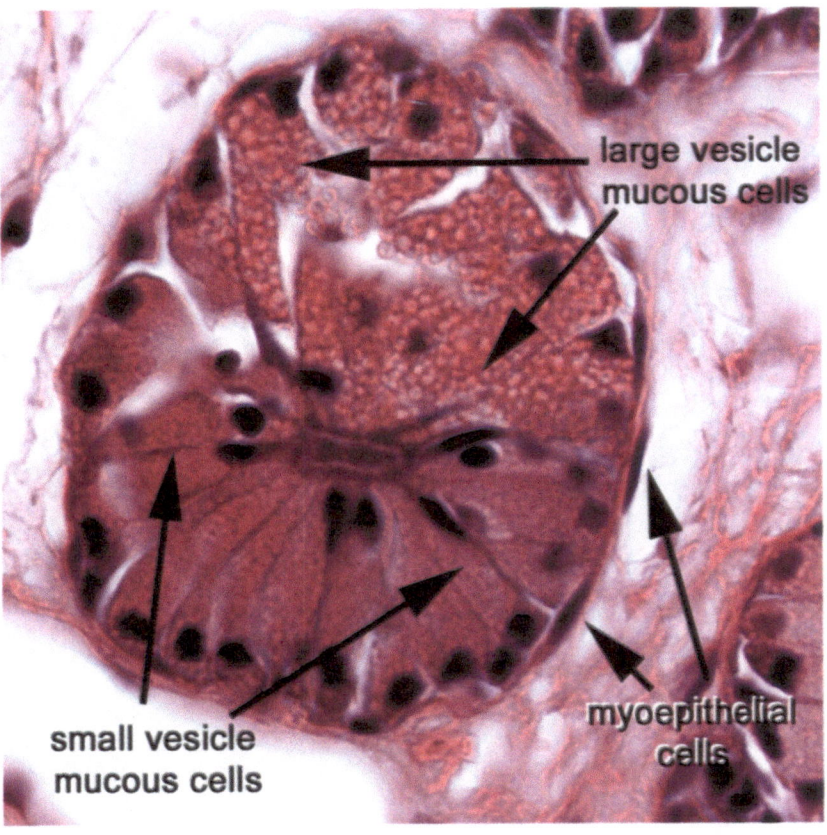

large vesicle
mucous cells

small vesicle
mucous cells

myoepithelial
cells

Figure 8. (ABOVE) High magnification image of salivary gland acinus. The acini contain mucous cells with either large or small vesicles in the apical portion of the cell. The nuclei are located basally. Nuclei close to the lumen of the acinus belong to ciliated cells. Myoepithelial cells surround the acinus and appear as cells with very elongated nuclei and cytoplasmic processes that extend long distances around the acinus. These latter cells contract, aiding in the expulsion of mucus from the acinus.

Figure 7. (LEFT) Salivary gland and duct. The salivary gland acini and duct contain two types of mucous cells, one of which contains small vesicles and one of which contains large vesicles. In addition, there are ciliated cells interspersed among the mucous cells.

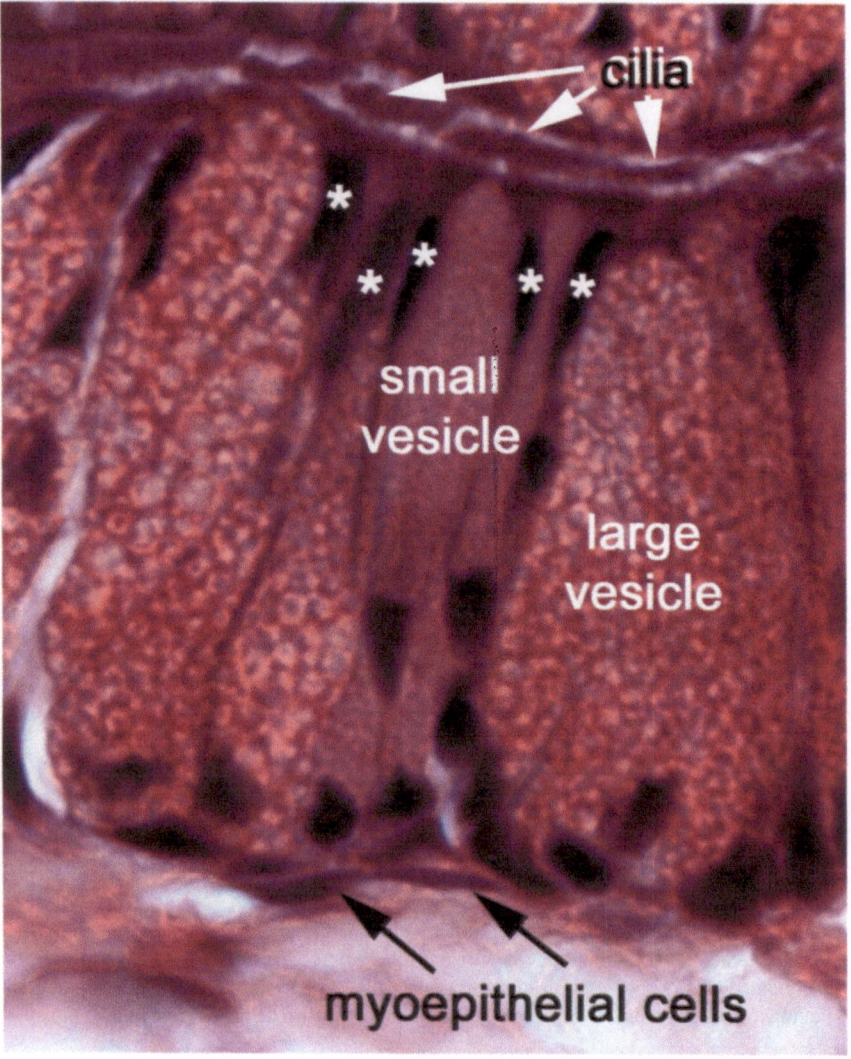

Figure 9. High magnification image of salivary gland cell types. Large vesicle cells are large wide cells whereas small vesicle cells are smaller and more narrow. Ciliated cells are very narrow and are squeezed between the mucous cells except for their apical end that fans out to provide a large apical surface with cilia. Their nuclei (asterisks) are wedged between the apical ends of the mucous cells. Myoepithelial cell nuclei are flattened against the base of the acinus.

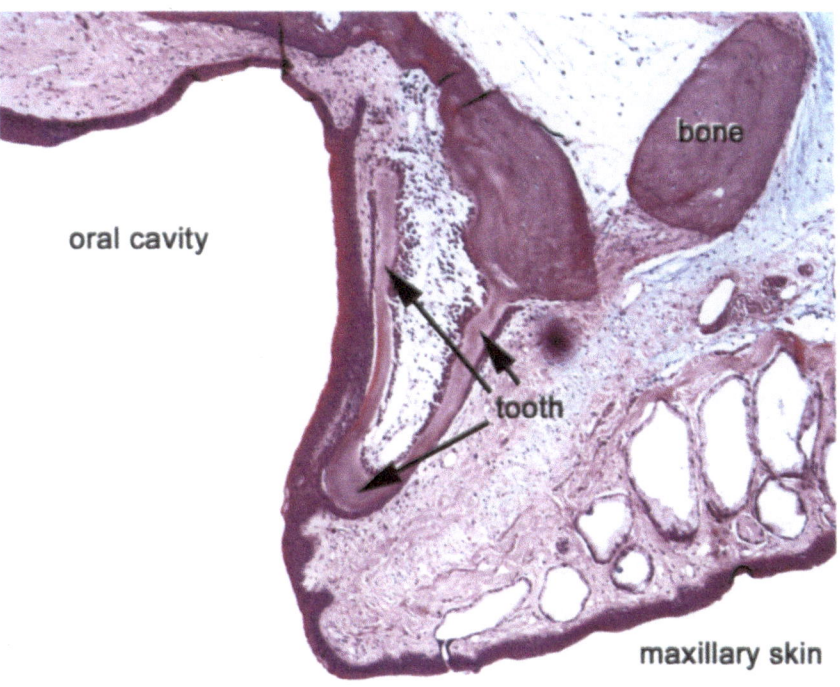

Figure 10. Structure and location of teeth. Teeth do not contain an outer layer of enamel. They are composed of dentin and are situated in the connective tissue underlying the oral epithelium. The rostral end of the tooth appears to be continuous with bone rather than being suspended in bone by a ligament.

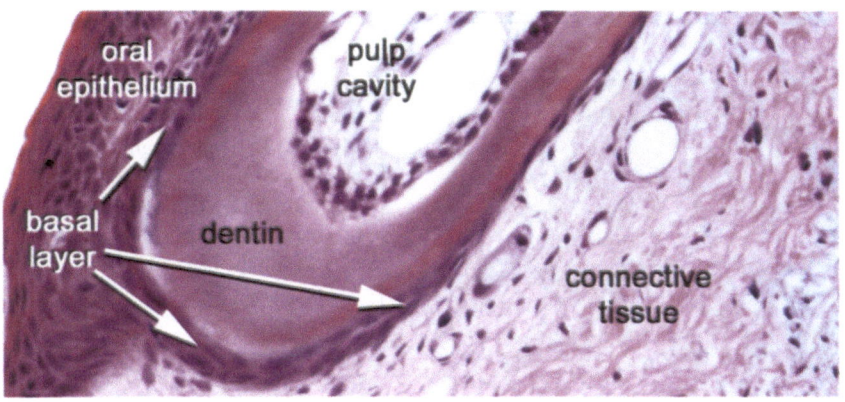

Figure 11. Magnified view of the tip or incisal edge of the tooth. The basal layer of oral mucosal epithelial cells surrounds the tooth.

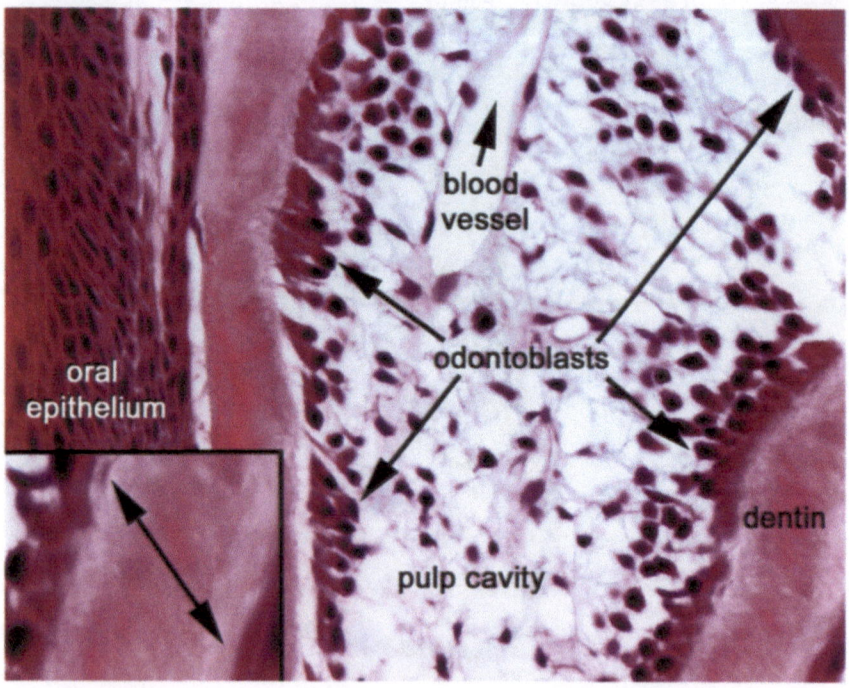

Figure 12. Magnified view of the tooth showing the odontoblast cells within the pulp cavity. Odontoblasts synthesize dentin, which is harder than bone. Within the dentin are tubules that contain the long processes of the odontoblasts. These tubules appear as striations through the dentin and are perpendicular to the surface of the dentin. **Inset** shows the orientation of the dentinal tubules through which the odontoblastic processes project.

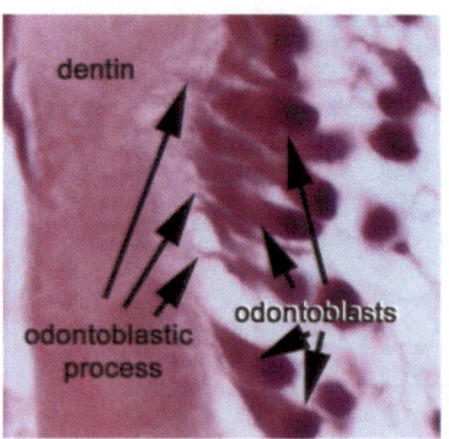

Figure 13. High magnification image of odontoblasts showing the projection of odontoblastic processes into the dentinal tubules within the layer of dentin.

III EPIGLOTTIS

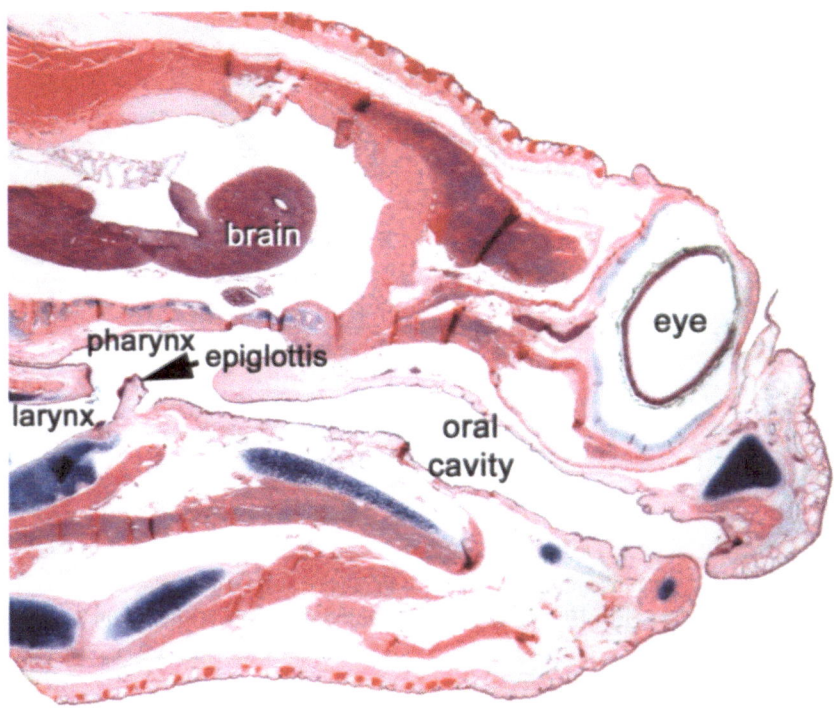

Figure 14. The epiglottis is a movable tissue that covers the larynx during swallowing preventing food from gaining access to the trachea. It is open (displaced in a rostral direction) during respiration.

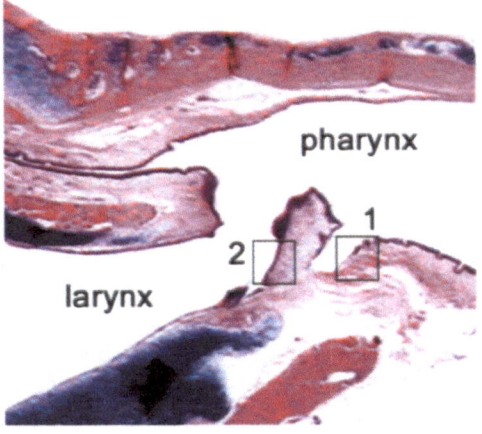

Figure 15. The epiglottis divides the regions of the pharynx and larynx. There is stratified columnar epithelium just rostral to the epiglottis (Box 1 and Figure 16 next page) and ciliated pseudostratified epithelium on the epiglottis (Box 2 and Figure 17 next page).

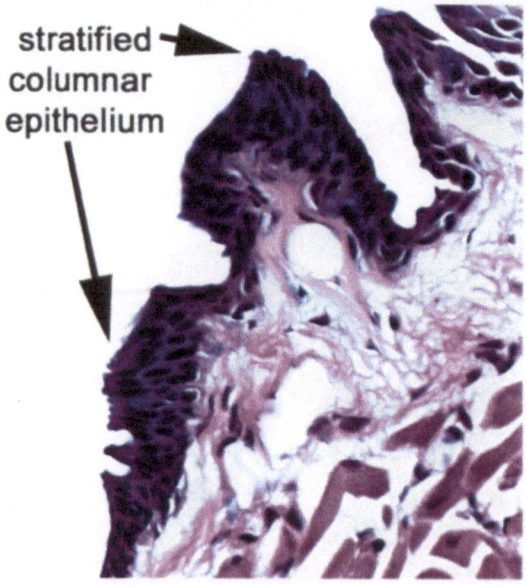

stratified
columnar
epithelium

Figure 16. High magnification image of the region just rostral to the epiglottis (Box 1 in Figure 15). This region contains a stratified columnar epithelium.

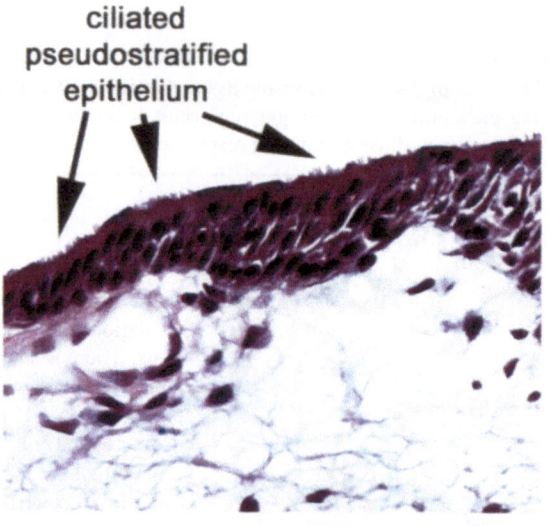

ciliated
pseudostratified
epithelium

Figure 17. High magnification image of the epithelium on the epiglottis (Box 2 in Figure 15). The epiglottis is covered by a ciliated pseudostratified epithelium.

IV ESOPHAGUS

The esophagus in *Xenopus* is composed of a nonkeratinized stratified squamous epithelium frequently invaginating into the underlying lamina propria. Deep to the connective tissue are inner circular and outer longitudinal muscularis externa layers.

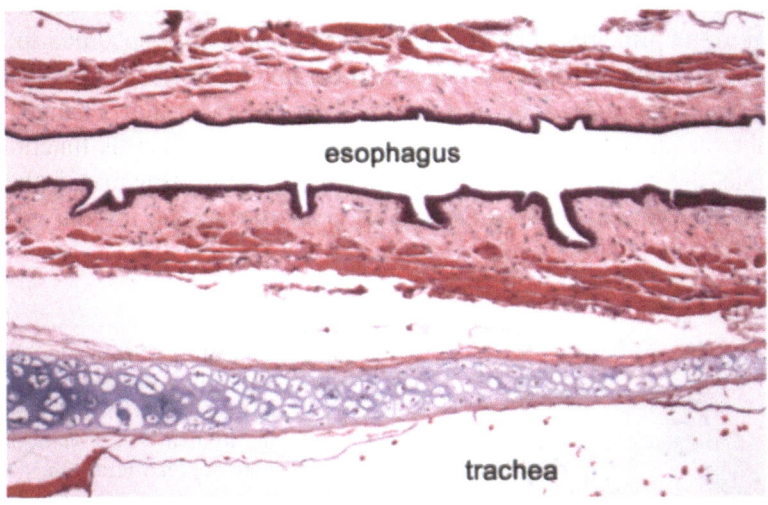

Figure 18. Low magnification images of the esophagus. The esophagus is dorsal to the trachea that is encased in cartilage.

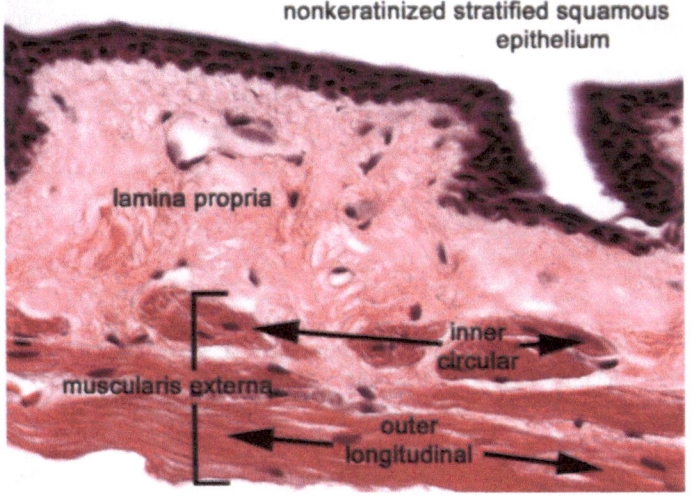

Figure 19. High magnification image of the esophagus. The esophagus is composed of stratified squamous epithelium, an underlying lamina propria and muscularis externa.

V NASAL CAVITY

Adult *Xenopus* possess three separate nasal chambers or cavities for the detection of odorants in air and water. Tadpoles possess two chambers, a principal chamber for detecting waterborne odorants and a vomeronasal organ that is thought to detect odors mainly from conspecific animals. During metamorphosis a third chamber, the middle chamber, develops rostral to the principal chamber. The latter chamber is re-specified to detect airborne odorants while the middle chamber takes over the role of detecting waterborne odorants. The vomeronasal organ does not change. Each chamber has unique histological characteristics that reflect its function. For example, the principal chamber has mucus-producing Bowman's glands in the lamina propria that secrete mucus onto the surface of the olfactory epithelium.

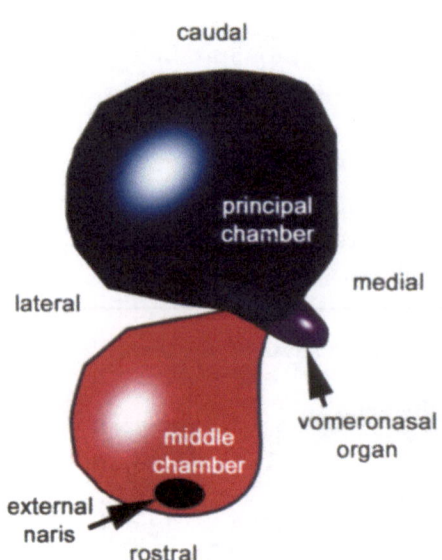

Figure 20. Schematic three-dimensional diagram of the chambers of the *Xenopus* nasal cavity from a dorsal view. This diagram does not reflect the actual shape of the chambers, which are rather convoluted, but rather the relative positions of the chambers. The middle chamber takes up a rostrolateral position. The principal chamber takes up a caudomedial position and the vomeronasal organ is a diverticulum off of the junction between the middle and principal chambers with the closed end projecting medially.

Figure 22. (RIGHT) Layers of chemosensory mucosa as exemplified by the principal chamber epithelium. Both olfactory and vomeronasal mucosa are composed of a pseudostratified epithelium and underlying lamina propria. Whereas the principal chamber and vomeronasal organ have glands, only the principal chamber demonstrates an obvious layer of mucus overlying the epithelium. The standard cell layers of epithelium from base to surface are a basal cell layer that gives rise to new cells, a sensory cell layer containing the cell bodies of chemosensory neurons and a sustentacular or supporting cell layer. Peripheral to this are the dendrites of the sensory cells intermingled with the processes of the supporting cells.

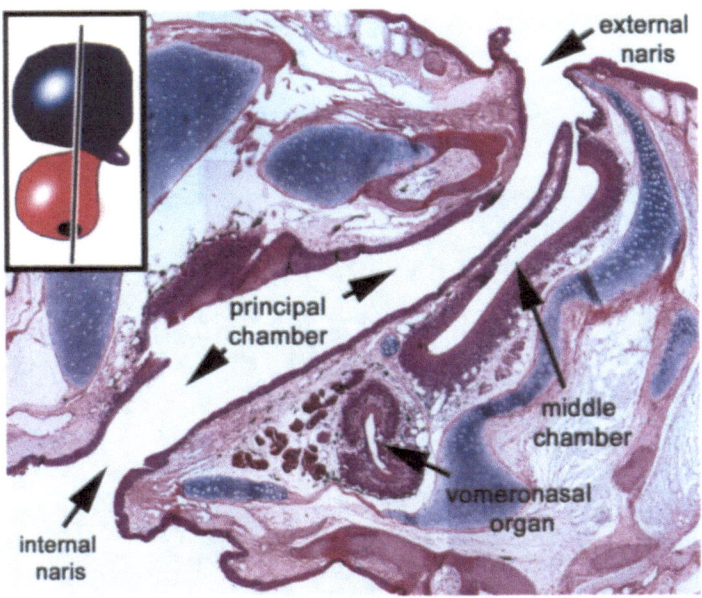

Figure 21. Microscopic image of three nasal chambers. **Inset** shows the plane of section in the parasagittal plane. The middle chamber is rostral (to the right), the principal cavity is caudal, and the vomeronasal organ is ventral to both other chambers. The principal cavity opens into the roof of the mouth as the internal naris.

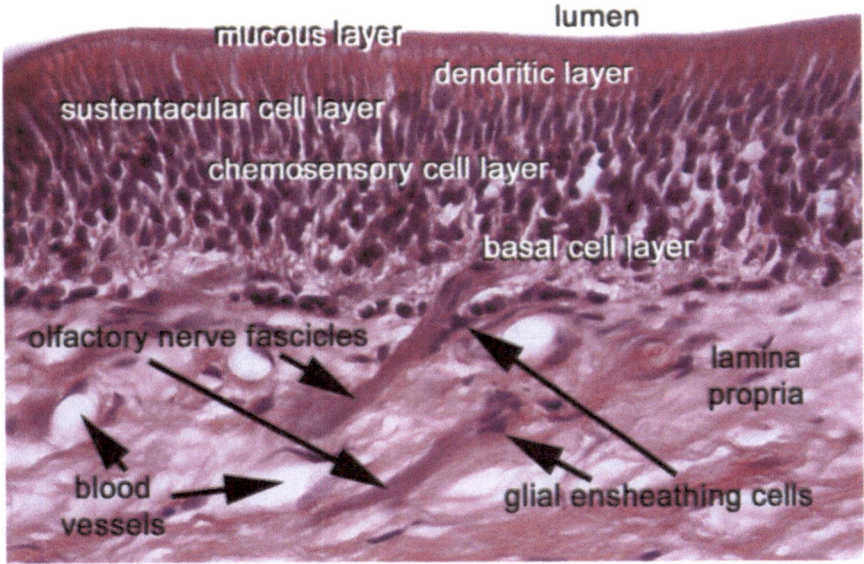

101

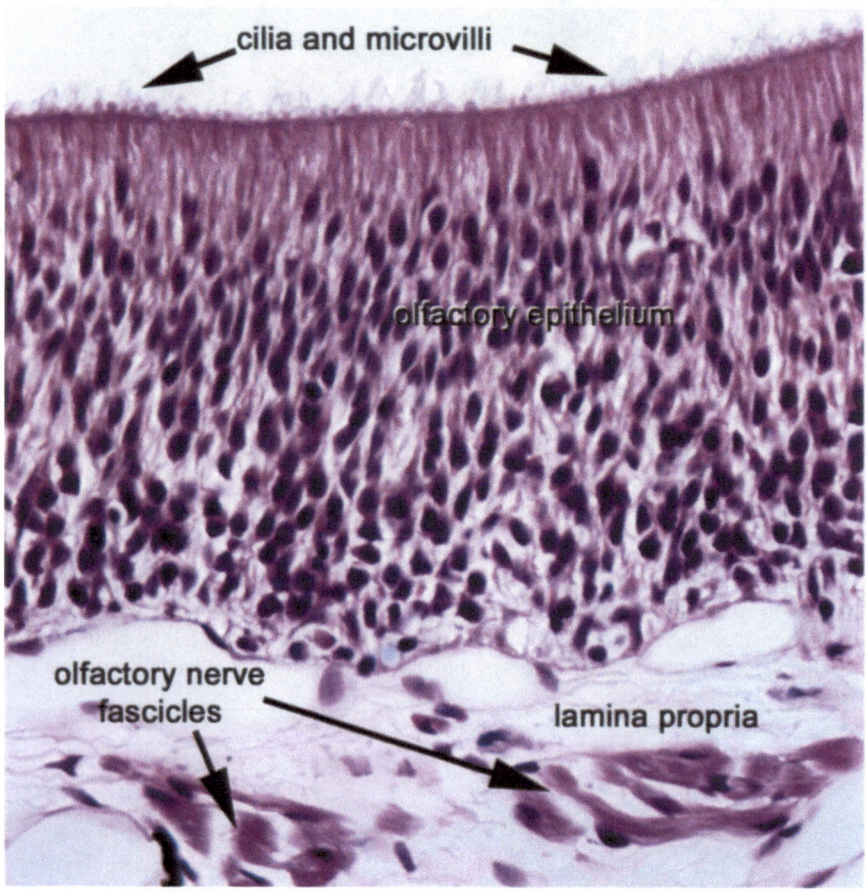

Figure 23. (ABOVE) High magnification image of the olfactory epithelium of the middle chamber. Note that there are no glands in the lamina propria. The olfactory epithelium of the middle chamber is composed of both ciliated and microvillar cells. This is unique in amphibians but similar to the olfactory epithelium of fishes.

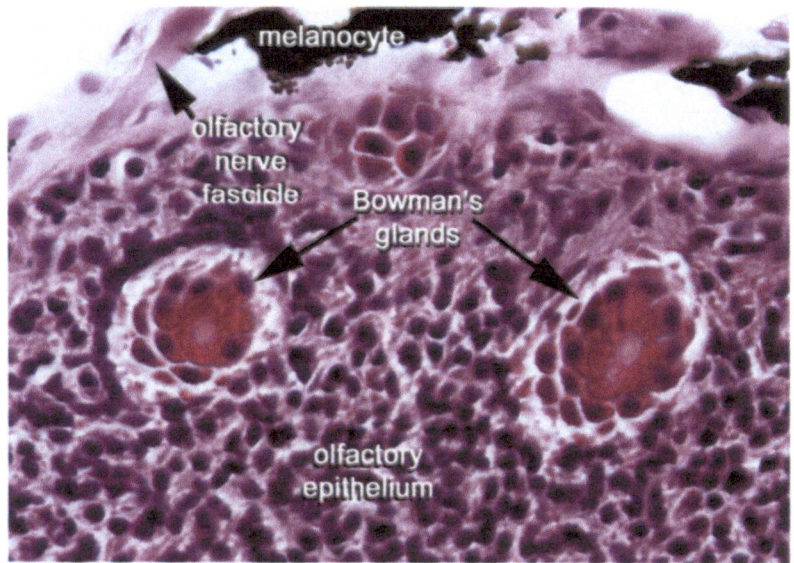

Figure 24. Tangential section of the olfactory epithelium in the principal chamber. Bowman's glands are cut in cross section demonstrating granules in the apical portion of the acinar cells.

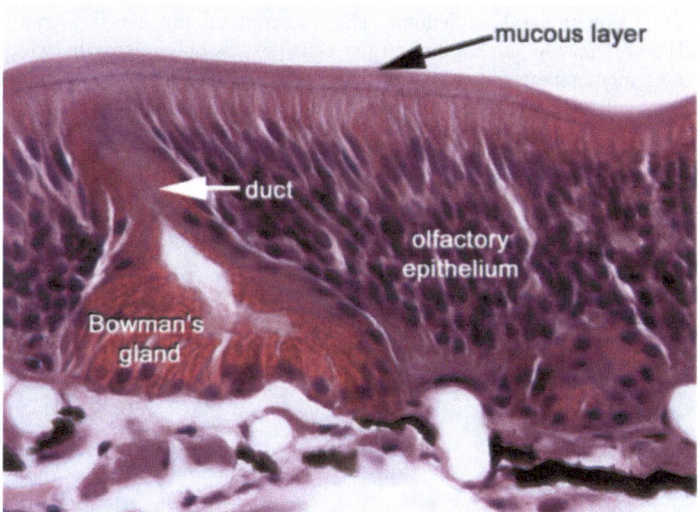

Figure 25. Transverse section of the olfactory epithelium in the principal chamber. Bowman's glands are oriented longitudinally demonstrating the acinar portion of the gland and the duct that carries mucus to the surface of the epithelium. The mucus layer can be seen overlying the dendritic layer of the epithelium. Cilia are embedded in the mucus but cannot be seen here

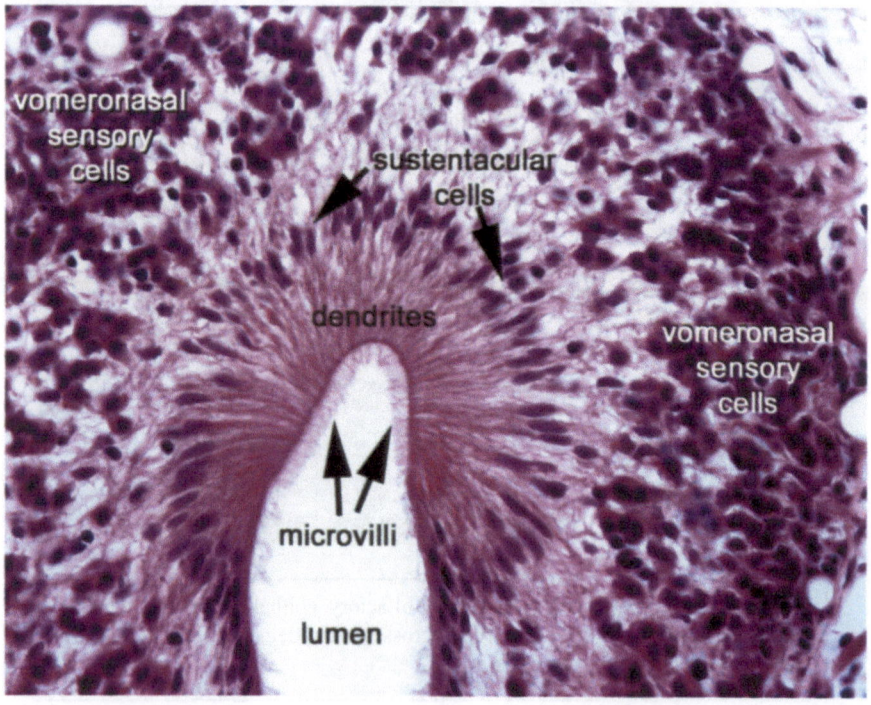

Figure 26. Vomeronasal epithelium. The vomeronasal mucosa is unique in that it surrounds a very narrow lumen, the sensory cells possess only microvilli, which are seen as very long membranous extensions of the sensory cell dendrite into the lumen, and the sustentacular cells occupy a very distinct layer separated from the sensory cells.

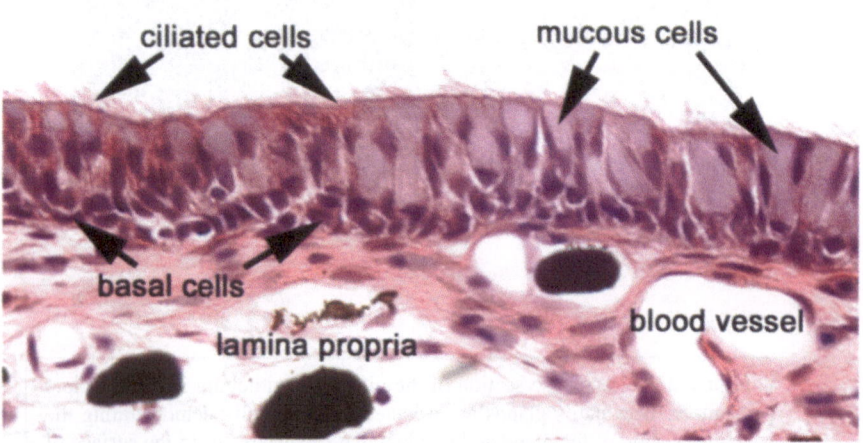

VI BRAIN

Xenopus frogs have a typical amphibian brain. Note that in all brain diagrams and micrographs, rostral is to the left.

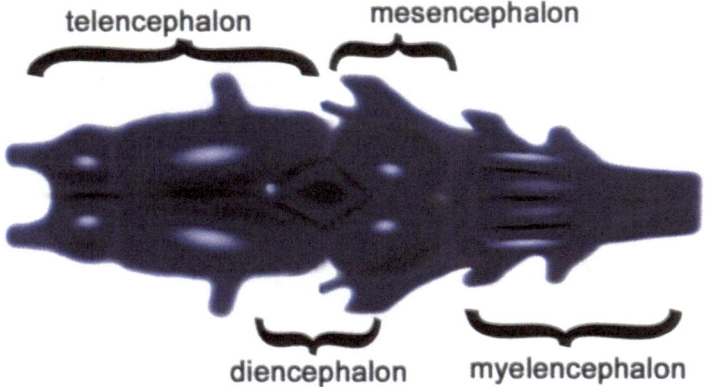

telencephalon   mesencephalon

diencephalon   myelencephalon

Figure 28. Diagram of the dorsal view of *Xenopus* brain showing the major brain divisions.

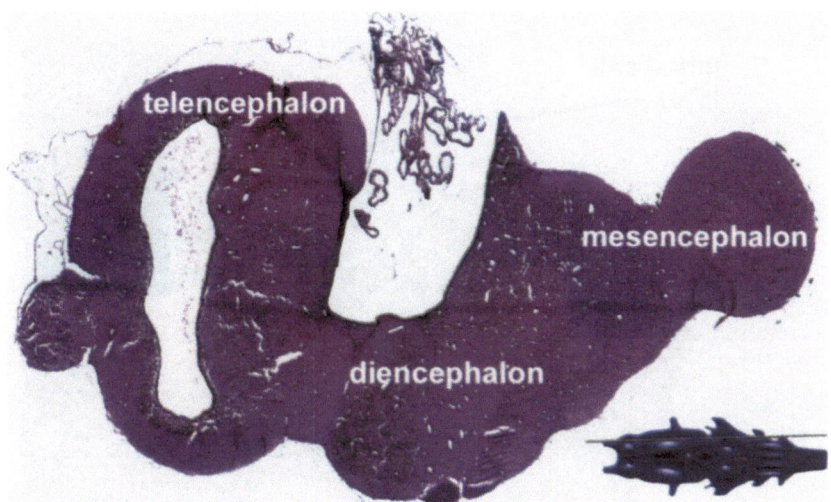

telencephalon

mesencephalon

diencephalon

Figure 29. Low magnification image of a parasagittal section of *Xenopus* brain showing three of the brain divisions. **Inset** shows plane of section.

Figure 27. (LEFT) High magnification image of respiratory mucosa from a ventrocaudal area of the nasal cavity just caudal to the vomeronasal organ. The respiratory epithelium contains numerous ciliated cells and mucous cells. Basal cells located at the base of the epithelium give rise to new respiratory cells. Underlying the epithelium is a typical lamina propria.

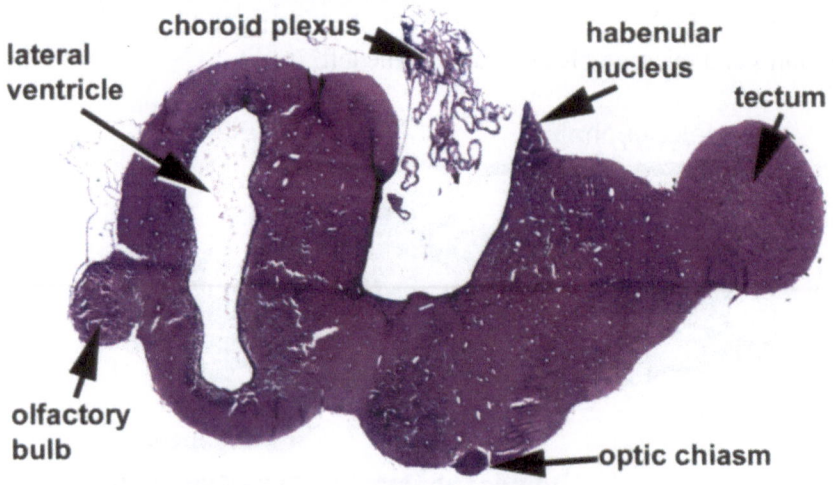

Figure 30. Parasagittal section of *Xenopus* brain demonstrating the location of several structures that will be illustrated at higher magnification below.

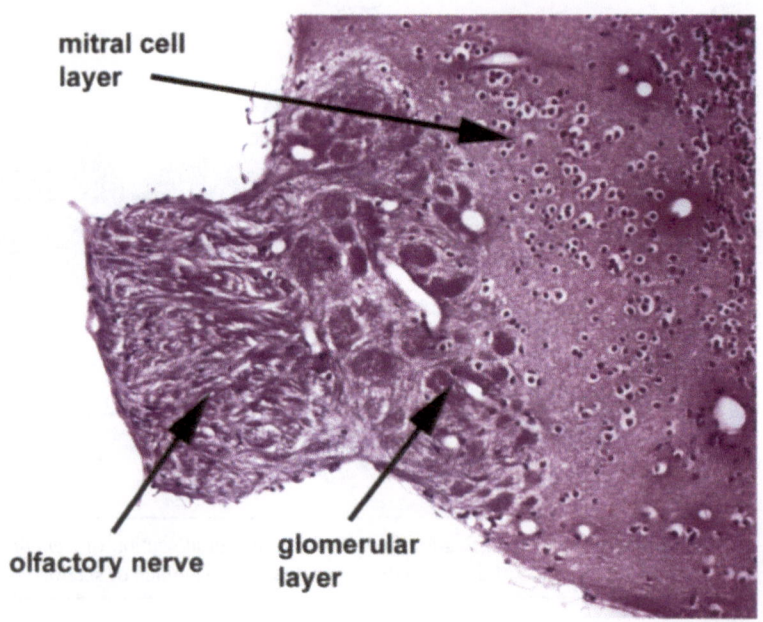

Figure 31. High magnification image of olfactory bulb showing the various layers. The olfactory nerve contains axons of the olfactory receptor neurons located in the nasal cavity. These axons project into the glomerular layer where they synapse with the dendrites of mitral cells. The cell bodies of the mitral cells are located in the mitral cell layer.

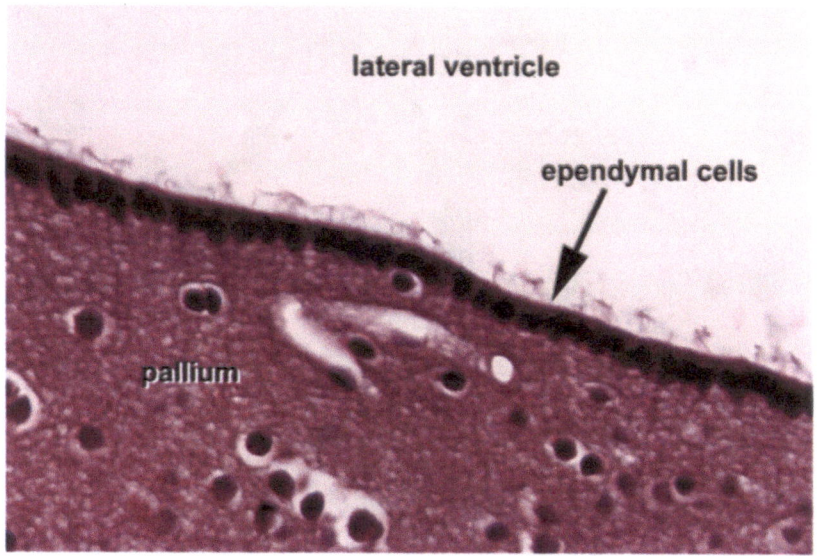

Figure 32. High magnification of the lateral brain ventricle and ependyma. The lateral ventricle is a space within the telencephalon that is filled with cerebral spinal fluid. The lateral ventricle is lined with ependymal cells that possess cilia. The underlying tissue in this micrograph is the neural tissue of the pallium, the layered structure of the telencephalon.

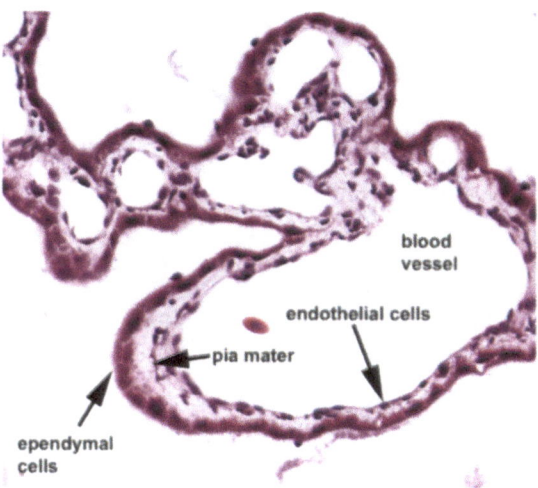

Figure 33. Low magnification image of the choroid plexus. The choroid plexus consists of an inner lining of ependymal cells and invaginating blood vessels with an intermedial layer of pia mater. Cerebral spinal fluid is produced by blood plasma passing through the pia mater and ependymal cells in the brain ventricles.

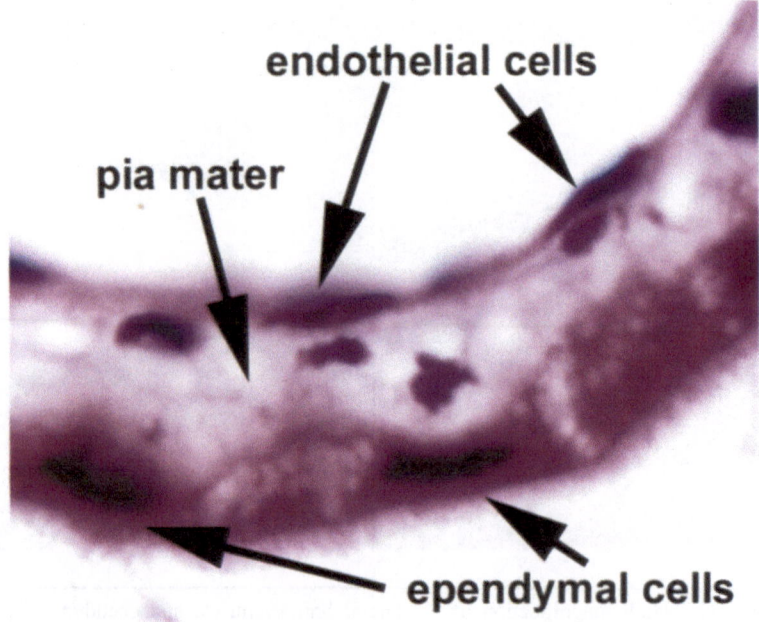

Figure 34. High magnification image of the choroid plexus illustrating the three layers: endothelial cells of the capillaries, pia mater and ependymal cells.

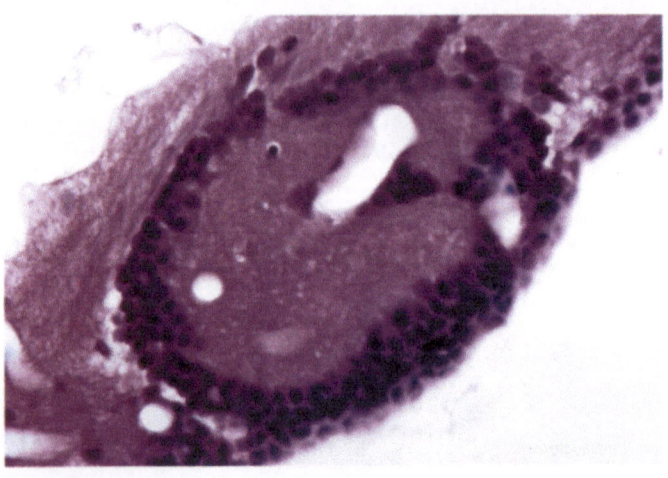

Figure 35. High magnification image of the habenular nucleus. This diencephalic nucleus has connections with the pineal gland that, in *Xenopus*, is embedded within the choroid plexus. In many amphibians and reptiles it is an asymmetric nucleus with two habenular nuclei on the left and one nucleus on the right side of the brain. In *Xenopus* habenular cell bodies are arranged around the rim of the nucleus whereas the neuropil in centrally located.

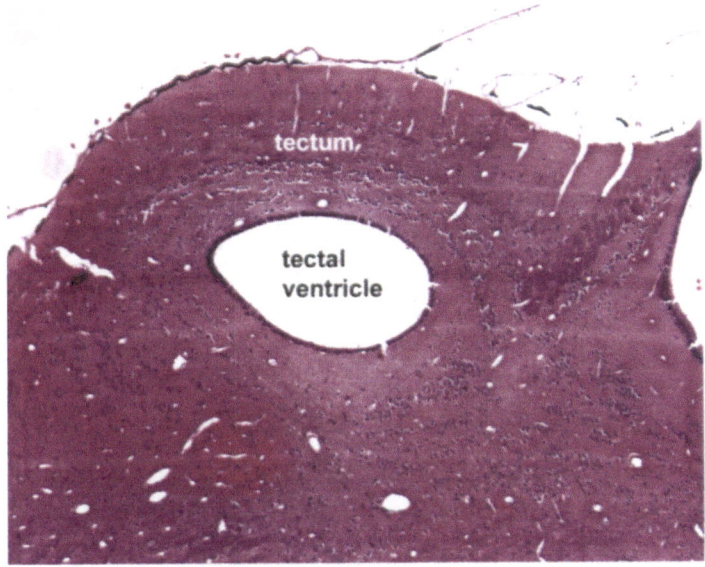

Figure 36. Low magnification image of the tectum. The tectum is a structure in the mesencephalon that receives heavy visual input. It is a layered cortical structure.

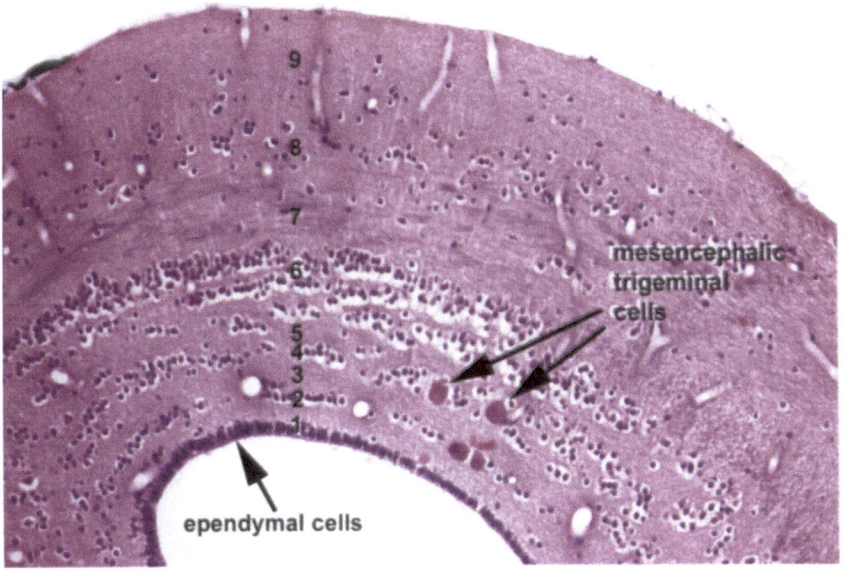

Figure 37. High magnification image of the tectal layers showing the mesencephalic trigeminal cells. The tectum has nine layers as indicated by the numbers. mesencephalic trigeminal cells are located in the deep layers. These cells respond to proprioceptive stimuli in the head region.

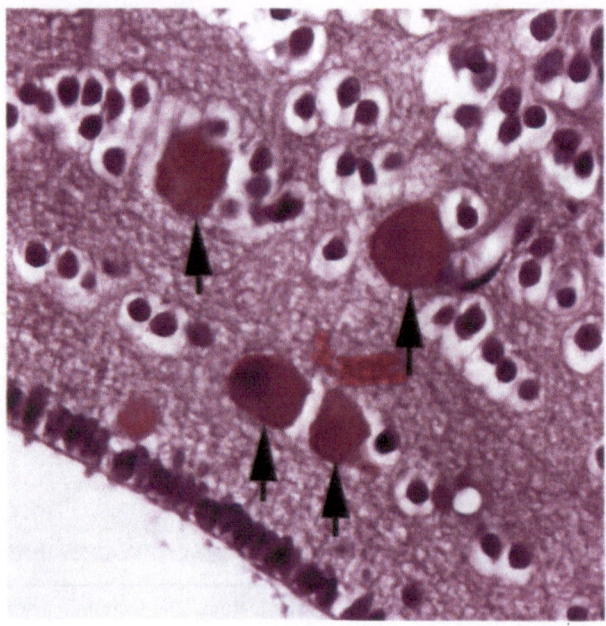

Figure 38. High magnification image of mesencephalic trigeminal cells (arrows).

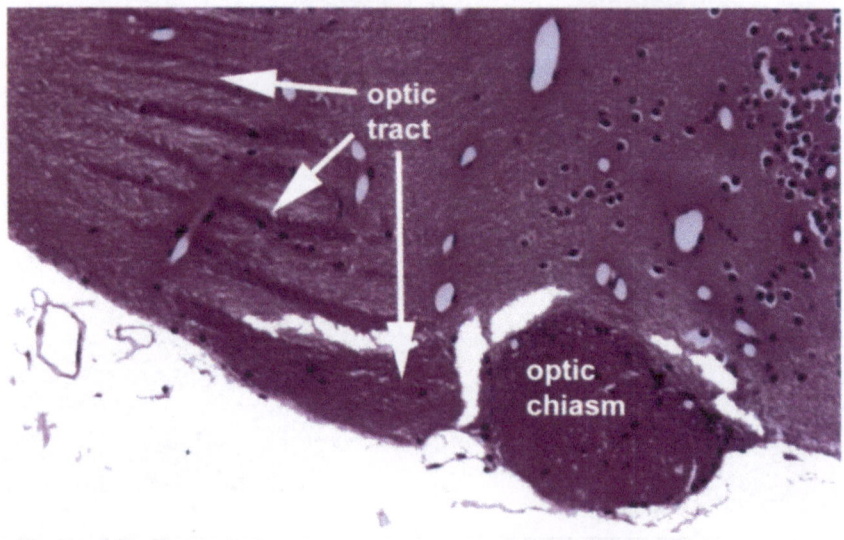

Figure 39. High magnification image of the optic chiasm and optic tract. The optic chiasm is the region in which axons from ganglion cells of the retina cross to project to the opposite side of the brain. These axons continue caudally in the optic tract.

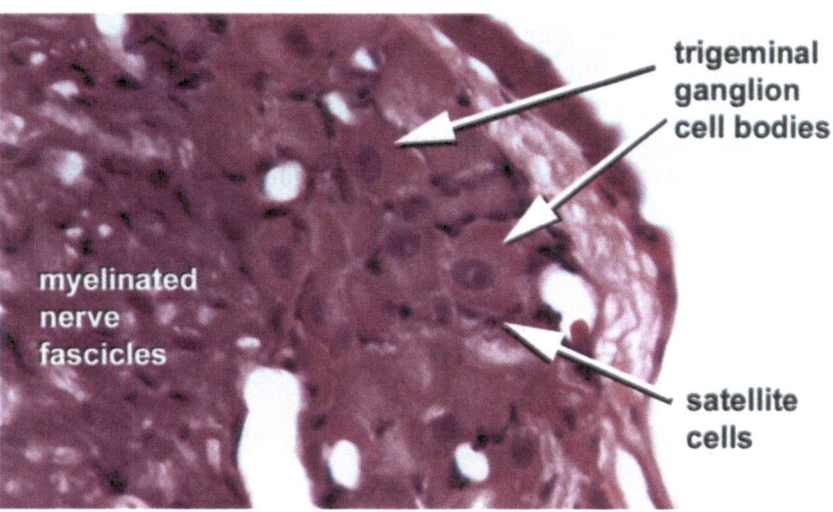

Figure 40. High magnification image of the trigeminal ganglion. The trigeminal ganglion is located ventral and slightly caudal to the optic chiasm. It is composed of clusters of large round somatosensory neurons with prominent nuclei and nucleoli. Myelinated processes are also grouped together in the ganglion. The cell bodies of the trigeminal neurons do not receive any synapses and are therefore surrounded by glial satellite cells.

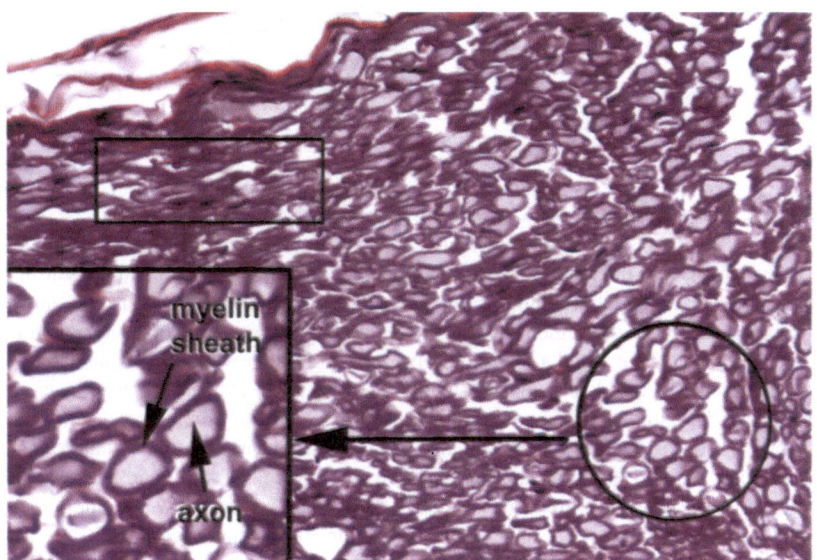

Figure 41. High magnification image of a myelinated nerve. Boxed area shows myelinated axons cut longitudinally and circle shows myelinated axons cut in cross-section. Inset shows details of myelin sheath and axonal compartment.

VII MIDDLE AND INNER EAR

The inner ear in *Xenopus* frogs contains the peripheral auditory and vestibular sensory structures. The arrangement of the various structures is shown in Figures 42 and 43 below. The auditory structures consist of the amphibian and basilar papillae. Having two auditory organs is unique to frogs and some salamanders. The vestibular structures include the maculus, sacculus, lagena and circular canals.

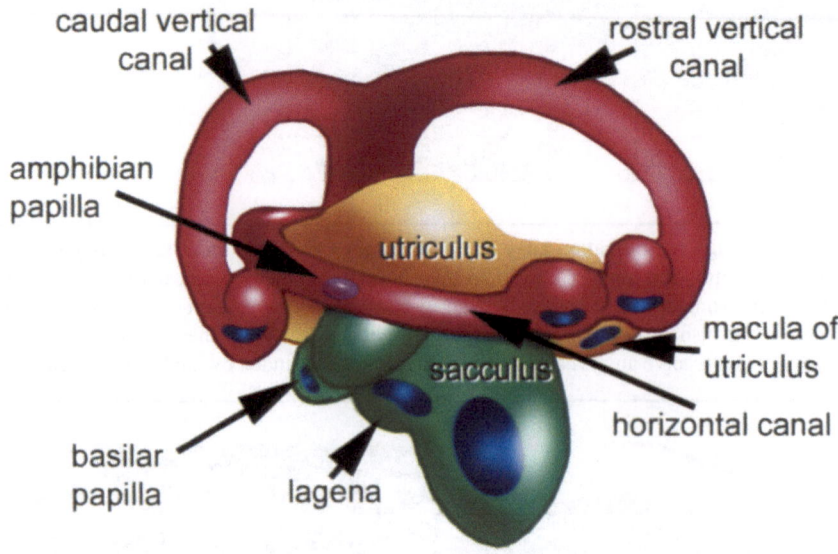

Figure 42. Lateral view of the right inner ear structures of a frog. The structure shown consists of the membraneous labyrinth containing endolymph and the sensory organs of audition and vestibular sense. The sensory portions of the organs are shown in blue. These are the regions which house the hair cells and received innervation from the VIII nerve ganglia. The auditory sensory structures are the amphibian and basilar papillae which are located within the sacculus. The basilar papilla is an evaginated region of the caudal wall of the sacculus. The amphibian papilla is a protuberance of the medial wall of the sacculus just ventral to the utriculus. The vestibular sensory structures are the cristae within the canals (red), and the maculae within the utriculus (yellow), the sacculus (green), and a ventral diverticulum off of the sacculus known as the lagena. The histological sections presented below are sections of the inner ear in the orientation shown above beginning in the lateral most parasagittal plane and proceeding through to the region of the basilar papilla and lagena that are located medially.

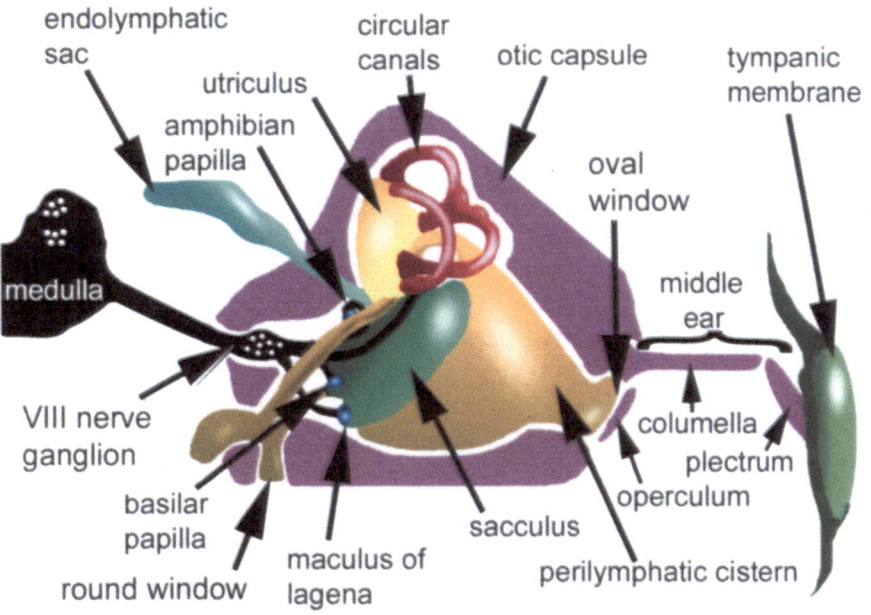

Figure 43. Caudal view of the outer (tympanic membrane), and middle and inner ear structures. The tympanic membrane is located on the outside of the animal. The middle ear ossicles, the plectrum, columella and operculum, transmit sound to the oval window. Pressure waves are set up in the perilymphatic system and are transmitted to the endolymphatic system which houses the sensory organs. All the sensory structures in the inner ear are innervated by the rostral and caudal VIII nerve and associated ganglia. The rostral ganglion innervates the horizontal and rostral vertical canals, the utriculus and a portion of the sacculus. The caudal ganglion innervates the caudal vertical canal, amphibian and basilar papillae, lagena and remaining portion of the sacculus.

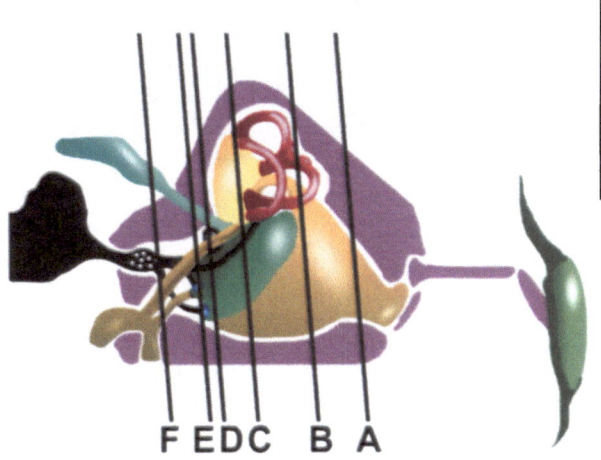

Figure 44. Diagram of ear in the coronal plane and as seen from a caudal view, showing planes of section for the figures in this chapter.

F E D C   B A

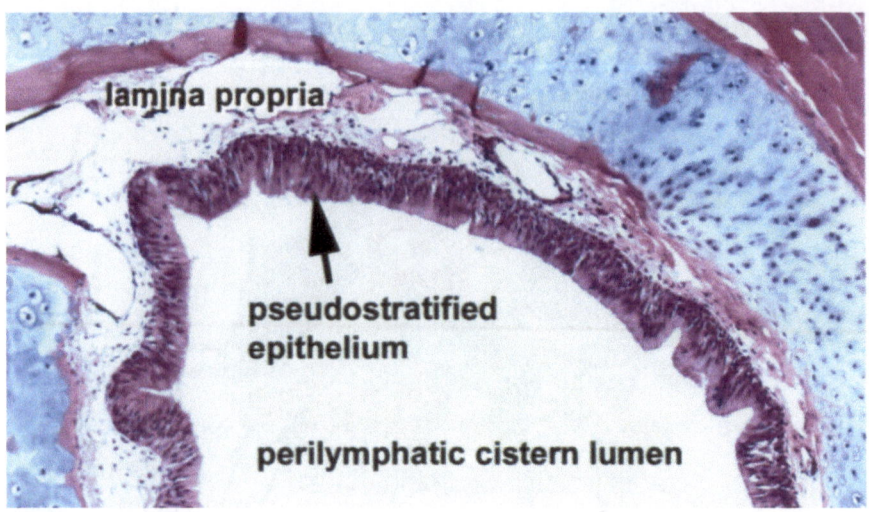

Figure 45. Low magnification image of the dorsal wall of the perilymphatic cistern (Plane A in Figure 44). The cistern is lined with a pseudostratified epithelium that appears identical to that of the respiratory epithelium of the nasal cavity.

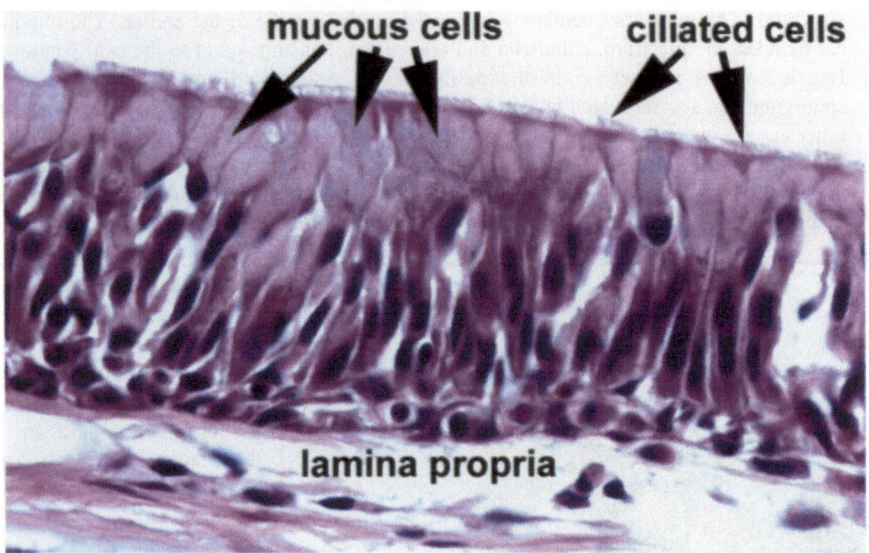

Figure 46. High magnification image of epithelium within perilymphatic cistern (Plane A in Figure 44). The epithelium is composed of ciliated cells and mucous cells. The ciliated cell apical ends are in the form of wedge shaped extensions with a broad surface membrane facing the lumen.

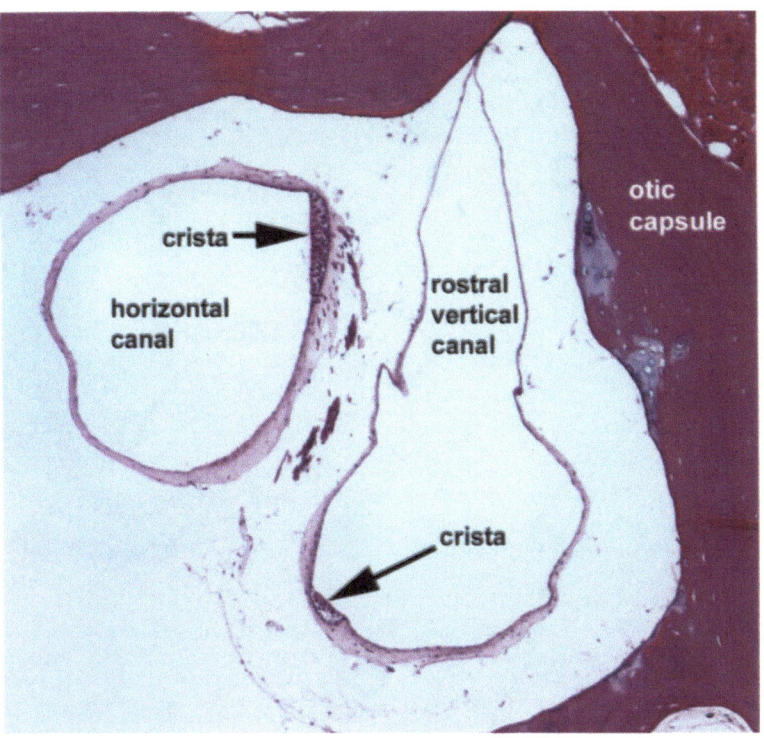

Figure 47. Low magnification image through two canals of the inner ear cistern (Plane B in Figure 44). The horizontal canal is oriented in cross-section and the rostral vertical canal is oriented obliquely. Cristae appear as crescents of sensory and supporting cells.

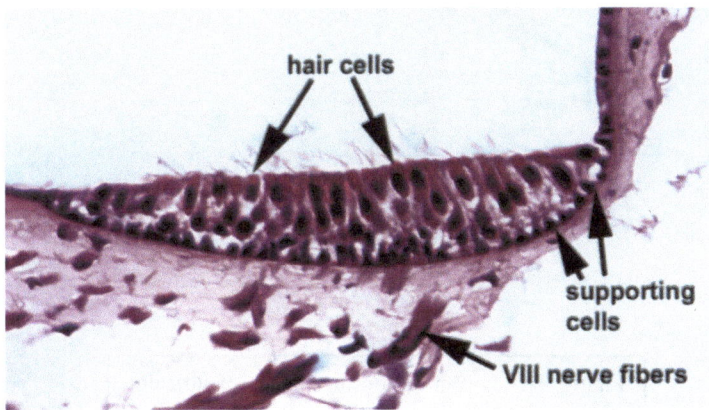

Figure 48. High magnification image of the cristae from the horizontal canal (Plane B in Figure 44). Image has been rotated counterclockwise 90 degrees.

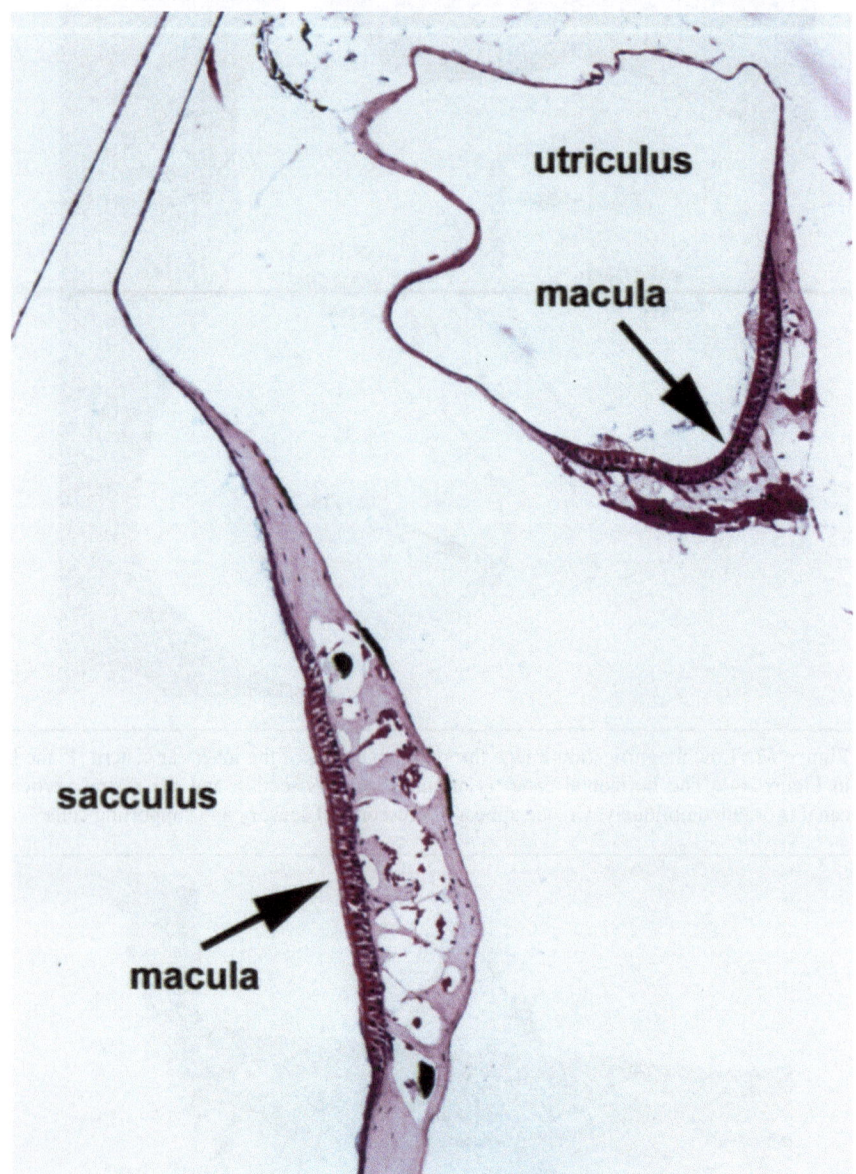

Figure 49. Low magnification image of utriculus and sacculus with macula (Plane C in Figure 44).

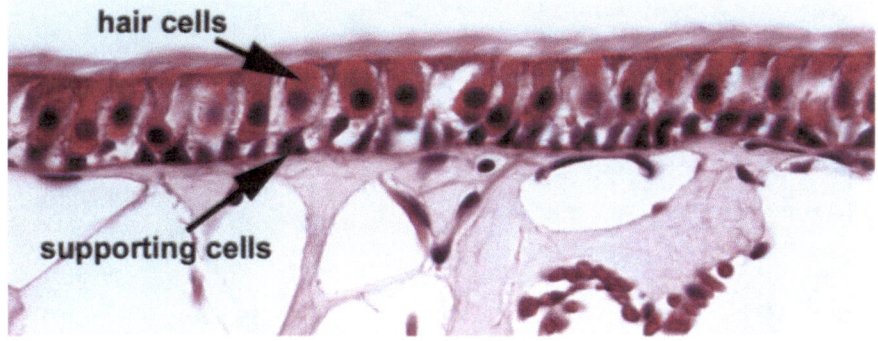

Figure 50. High magnification of macula of the sacculus (Plane C in Figure 44). The macula of the sacculus is a large organ composed of sensory hair cells and supporting cells.

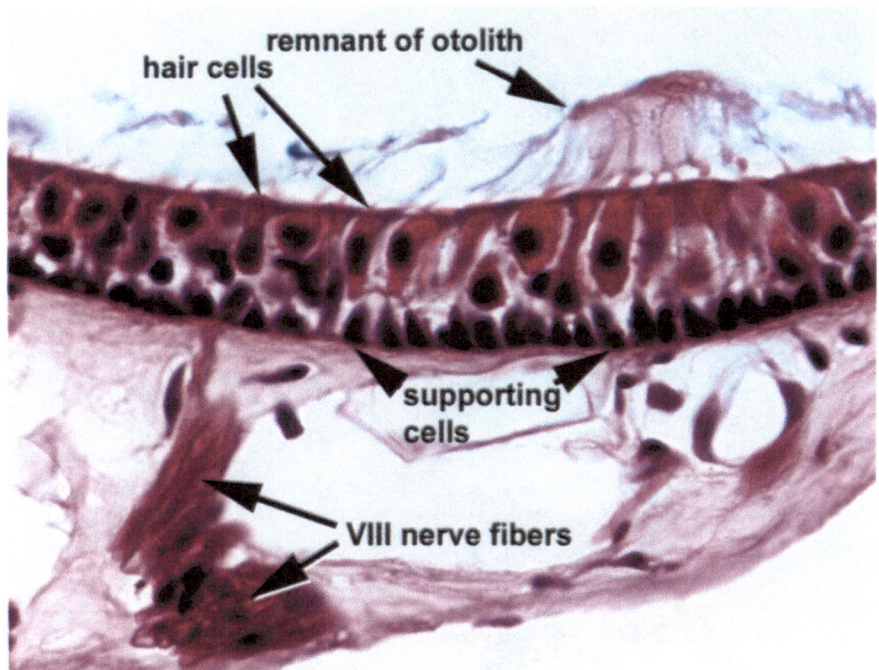

Figure 51. High magnification image of the macula of the utriculus. This organ is crescent-shaped in cross-section (only the rostral wall is shown here). Remnants of the otolithic membrane are seen over the hair cells. Nerve fibers of the VIII nerve can be seen coursing to the organ.

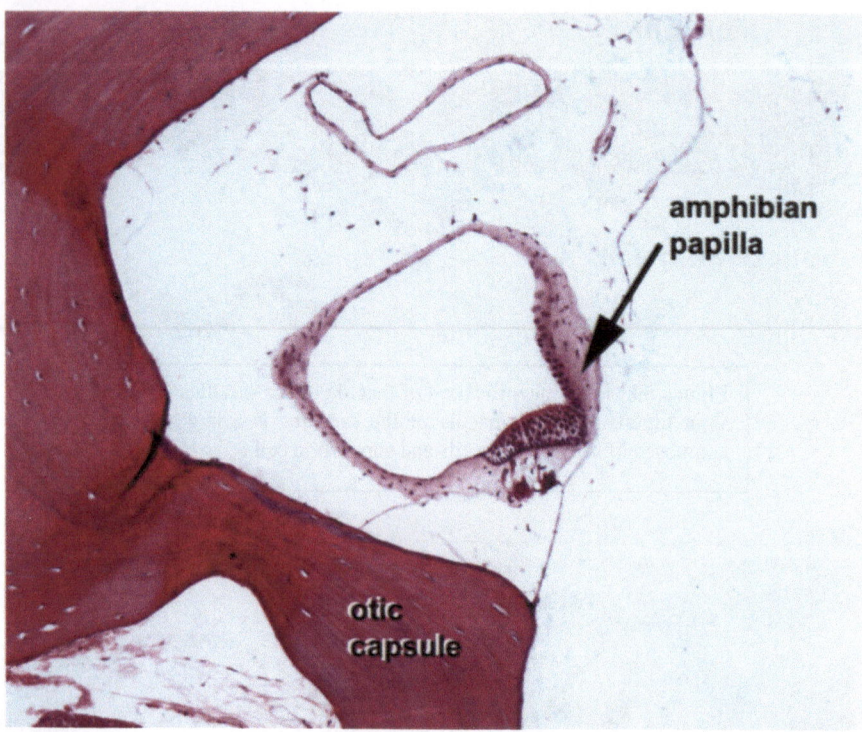

Figure 52. Low magnification image of the amphibian papilla (Plane D in Figure 44). This organ is located on the caudal wall of the sacculus. It is an auditory structure and therefore responds to vibrations in the air or water.

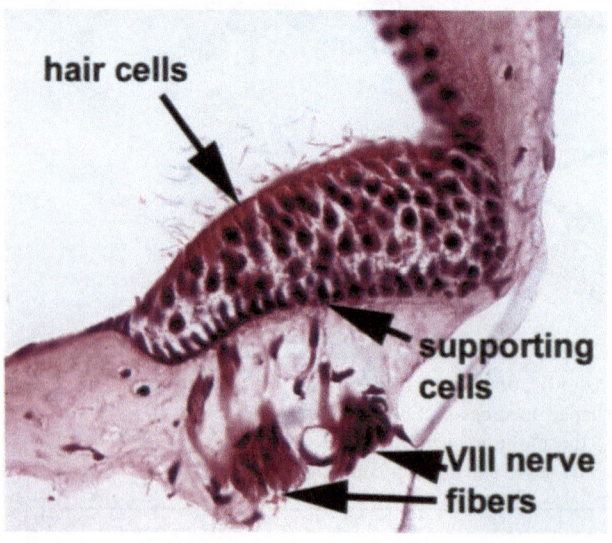

Figure 53. High magnification image of the amphibian papilla.

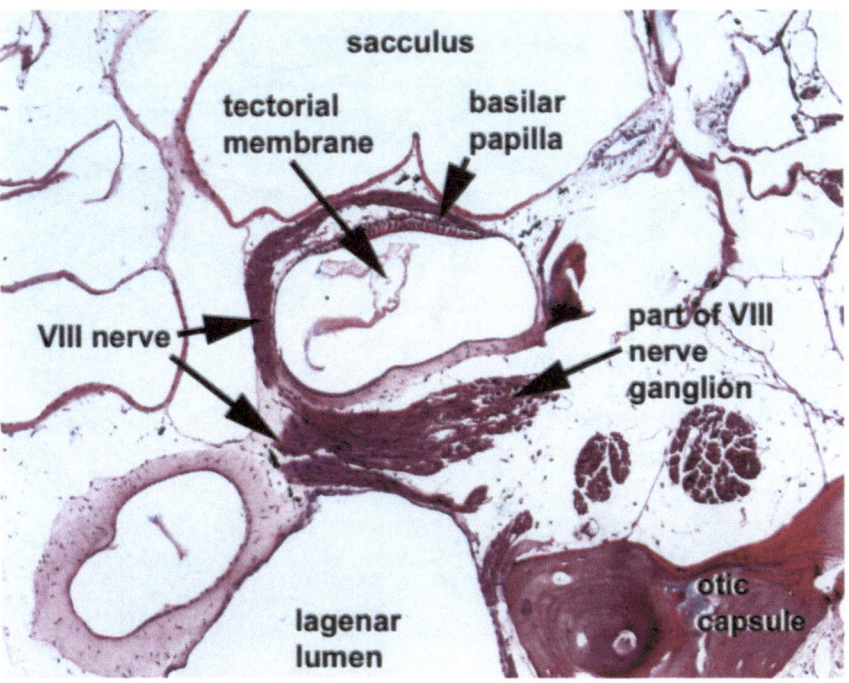

Figure 54. Low magnification image of the basilar papilla (Plane E in Figure 44). This is an auditory structure similar to the amphibian papilla. Note the tectorial membrane. This membrane moves with the movement of the endolymph and stimulates the hair cells of the basilar papilla. As seen from this parasagittal section, the basilar papilla is just dorsal to the lagena, which is a vestibular organ.

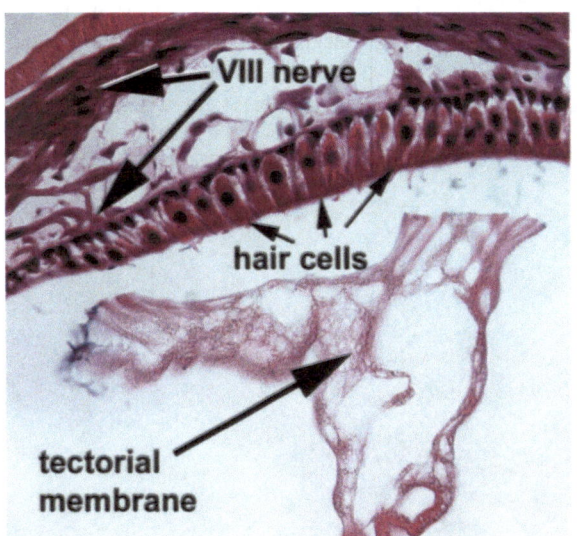

Figure 55. High magnification image of the basilar papilla. The tectorial membrane normally rests on the apical surface of the hair cells. VIII nerve fibers can be seen projecting to the basal surface of the organ to innervate the hair cells.

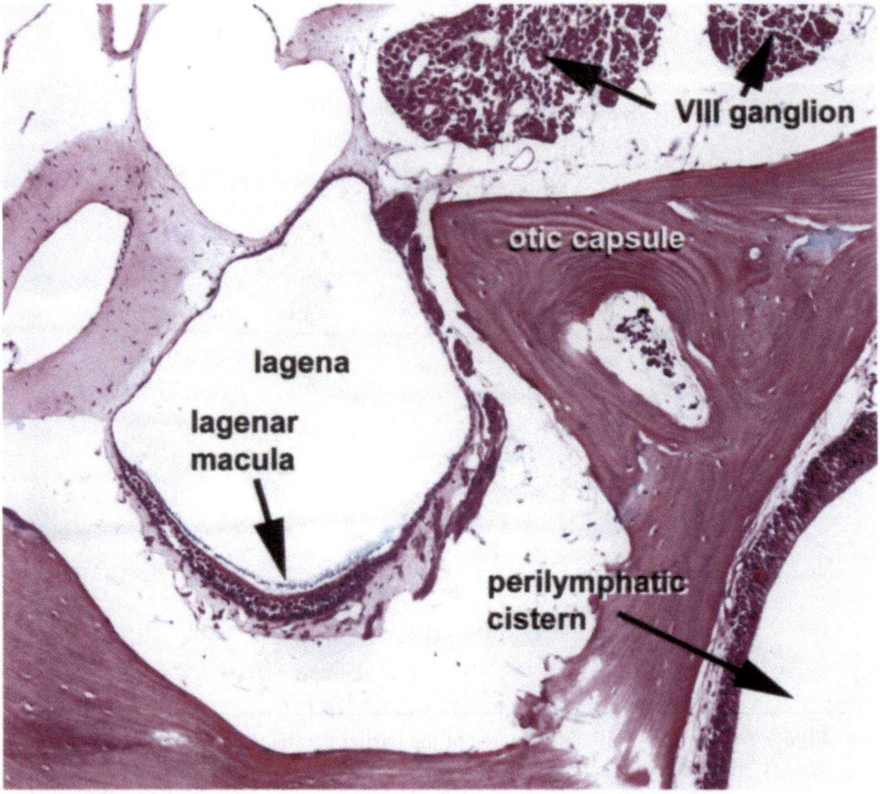

VIII ganglion

otic capsule

lagena

lagenar macula

perilymphatic cistern

Figure 56. (ABOVE) Low magnification image of the lagena (Plane E in Figure 44). The lagena is a vestibular organ. The macula is covered by an otolithic membrane which appears here as fuzzy blue material over the hair cells of the macula.

Figure 57. (BELOW) High magnification image of the lagenar macula.

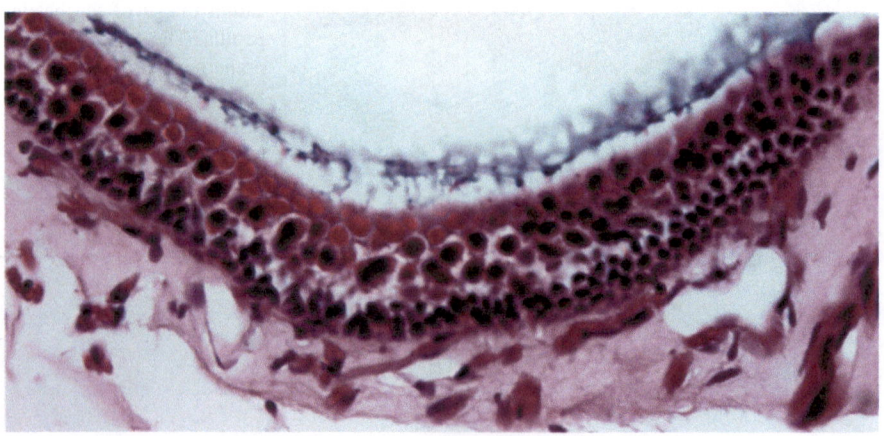

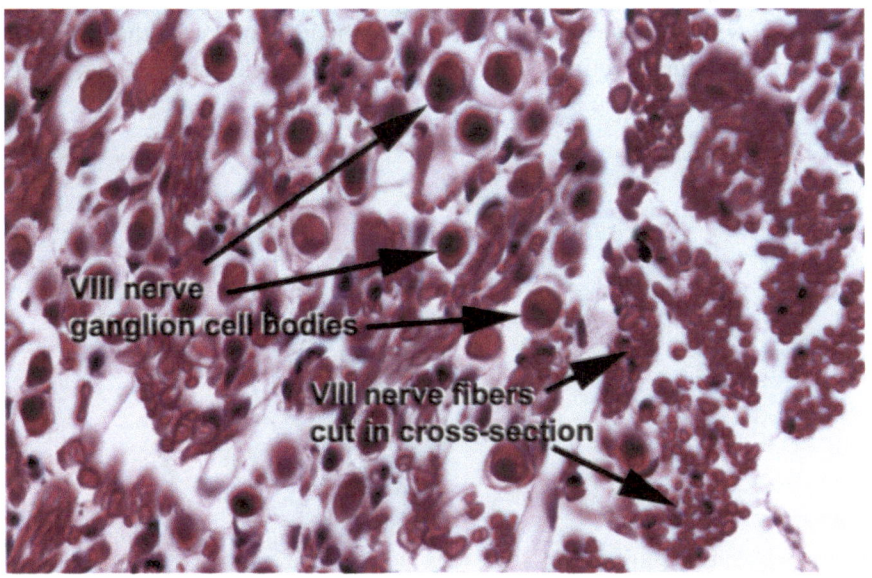

Figure 58. High magnification image of the VIII nerve ganglion (Planes E and F in Figure 44). Cell bodies of the VIII nerve sensory neurons appear as small round cell bodies with a somewhat eccentric nucleus. Myelinated axons travel in bundles. These cells innervate the hair cells of the auditory and vestibular ganglia.

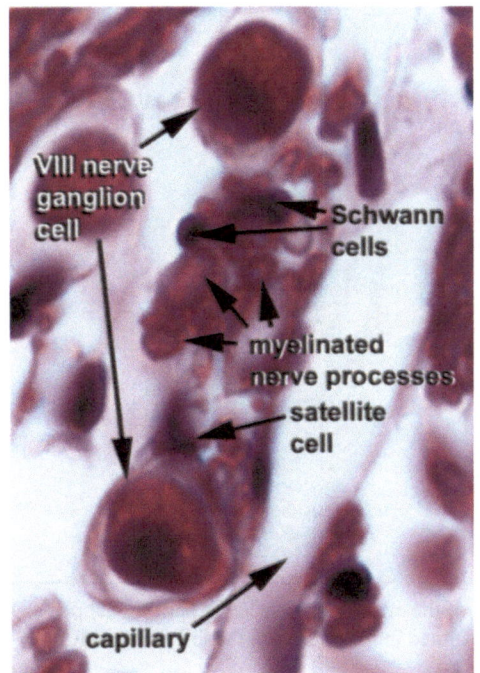

Figure 59. High magnification image of VIII nerve ganglion cells and their glial ensheathing satellite cells, and ganglion cell processes with their glial ensheathing Schwann cells. Schwann cells (small dark purple nuclei) among nerve processes) provide a myelin sheath to the nerve processes (here appearing as a purple circle around a clear center where the axon resides). VIII ganglion cells demonstrate a round nucleus with a prominent nucleolus.

VIII EYE AND ASSOCIATED STRUCTURES

Xenopus frogs have rather small eyes compared to many other frogs but have well-developed retinas. The sclera contains a cartilaginous layer that adds extra support to the eye.

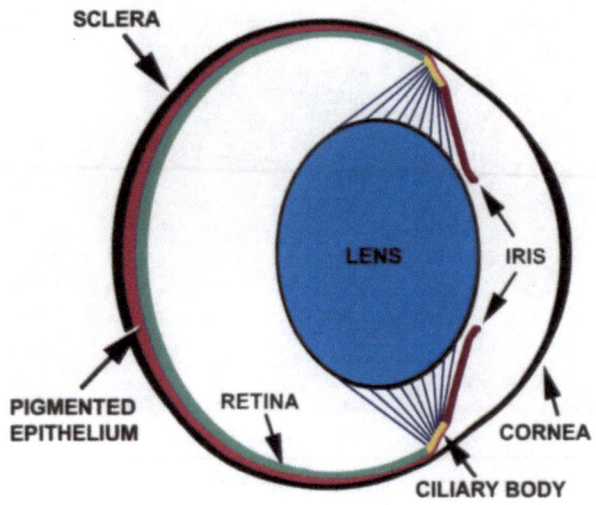

Figure 60. Diagram of the eye of *Xenopus laevis*.

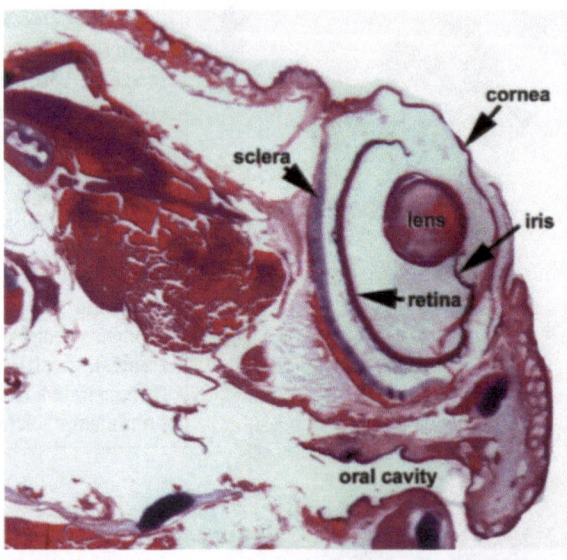

Figure 61. Low magnification image of the eye of *Xenopus* in the parasagittal plane. Details of the ocular structures are illustrated in the figures on the following pages.

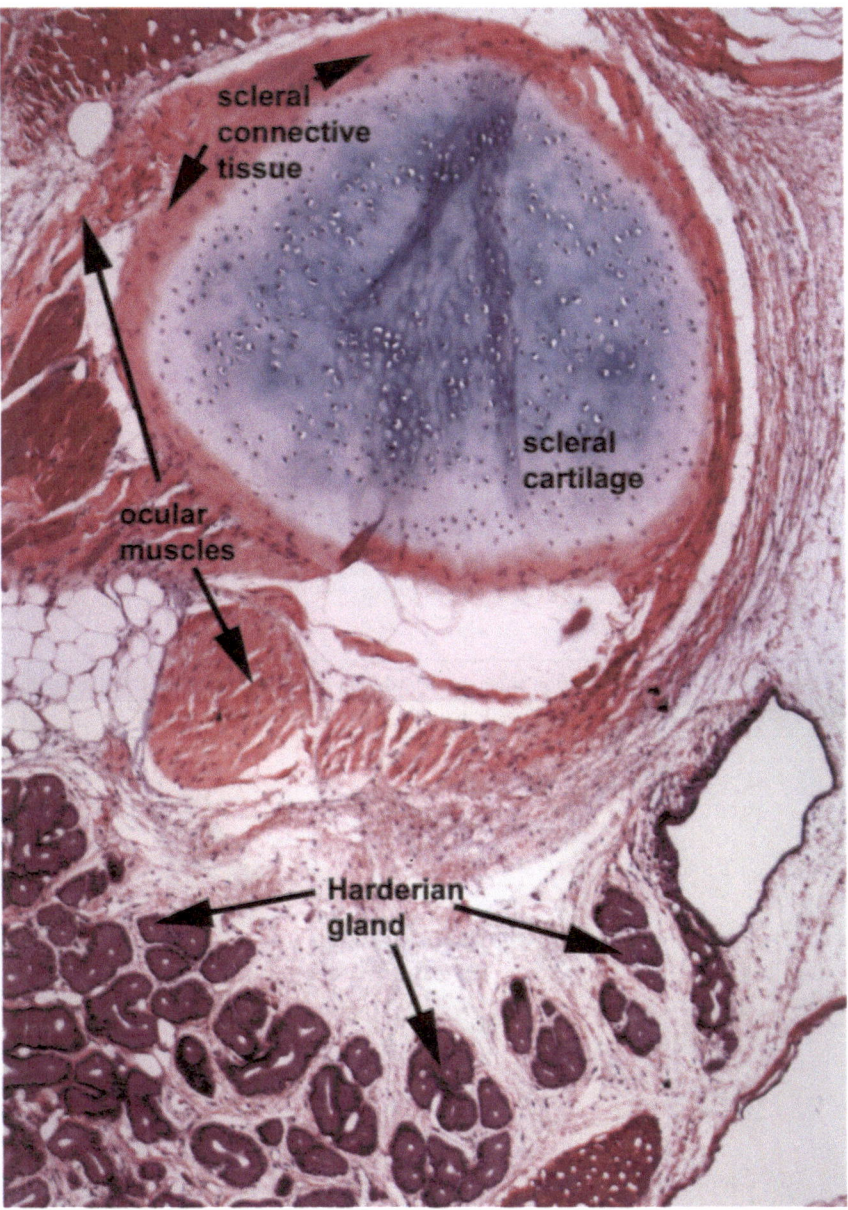

Figure 62. Low magnification image of the sclera, ocular muscles and Harderian gland. The ocular muscles attach to the sclera. The sclera contains a layer of cartilage for added support to the eye. The Harderian gland is located within the orbit and empties via a duct onto the cornea. In addition, Harderian gland secretions are thought to gain access to olfactory structures via the nasolacrimal duct and may play a role in chemosensory function.

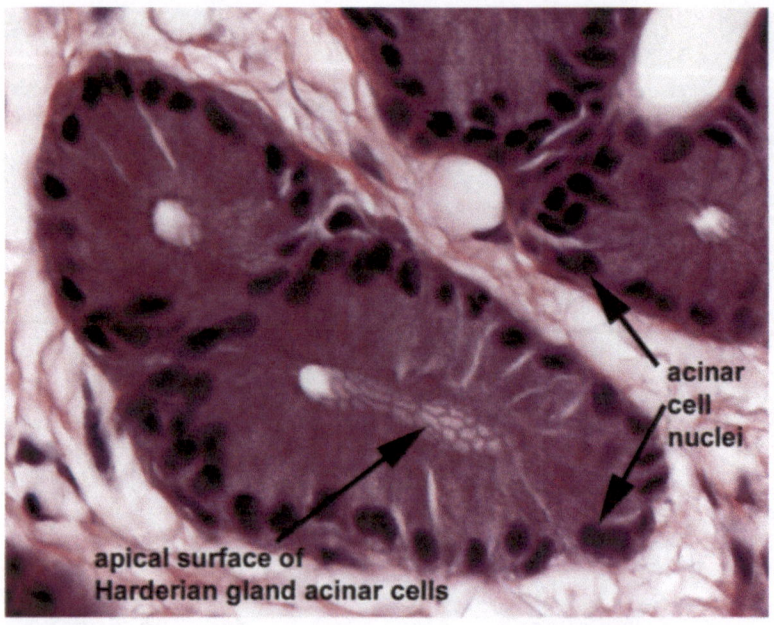

Figure 63. High magnification image of Harderian gland. At the apical surfaces of the glandular cells, secretions are expelled into the Harderian gland lumen.

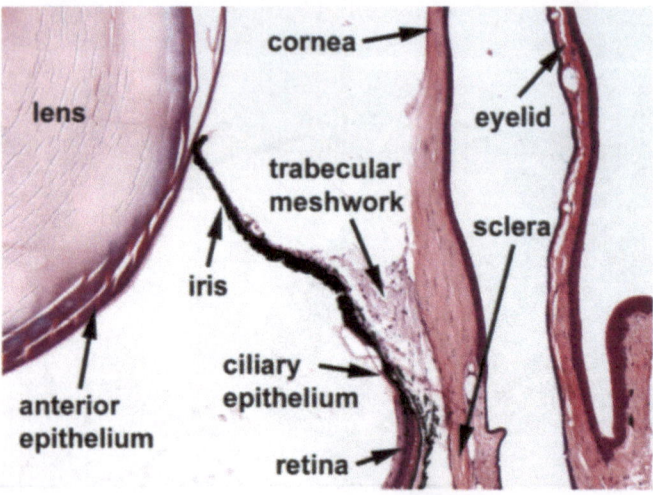

Figure 64. Low magnification image of the anterior segment. The ciliary epithelium of the ciliary body is continuous with the retinal pigment epithelium, and also with the epithelium of the iris. The dense connective tissue of the sclera is continuous with the transparent corneal stroma. Deep to the lens capsule are a few layers of epithelium that give rise to lens fibers.

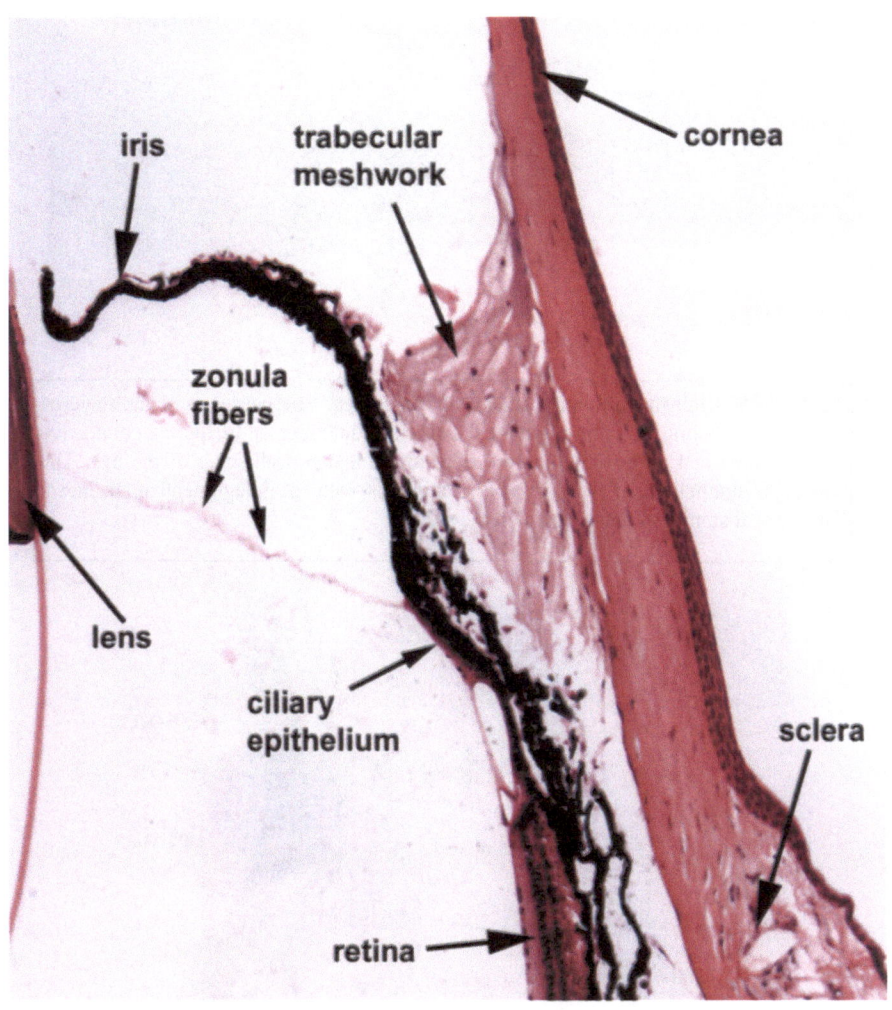

Figure 65. Iris, cornea, and ciliary body. The ciliary epithelium is comprised of two layers of simple cuboidal epithelium, with the apical surfaces of both layers in direct contact with each other. The inner layer is non-pigmented, and the outer layer (which is continuous with the retinal pigment epithelium) is pigmented. The two layers of epithelium in the iris are arranged in the same manner as in the ciliary epithelium, except that both layers are pigmented. Thin zonular fibers are firmly attached to the basal layers of the non-pigmented ciliary epithelium, and extend and attach to the lens capsule. Contraction of smooth muscle in the ciliary body allows for lens accommodation. Aqueous humor is secreted by the ciliary epithelium, and flows from the vitreal chamber to the anterior chamber. The aqueous fluid leaves the eye through the trabecular meshwork and canal of Schlemm to enter episcleral veins.

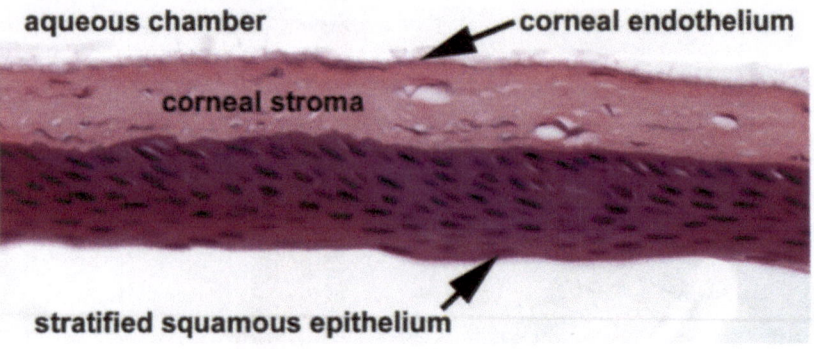

aqueous chamber   **corneal endothelium**

corneal stroma

stratified squamous epithelium

Figure 66. High magnification image of the cornea. The cornea is a multilayered structure consisting of a non-pigmented stratified squamous epithelium, a connective tissue stroma and an endothelium adjacent to the aqueous chamber of the eye. The squamous epithelium of the cornea is continuous with the integument of the head. The corneal stoma is continuous with the sclera.

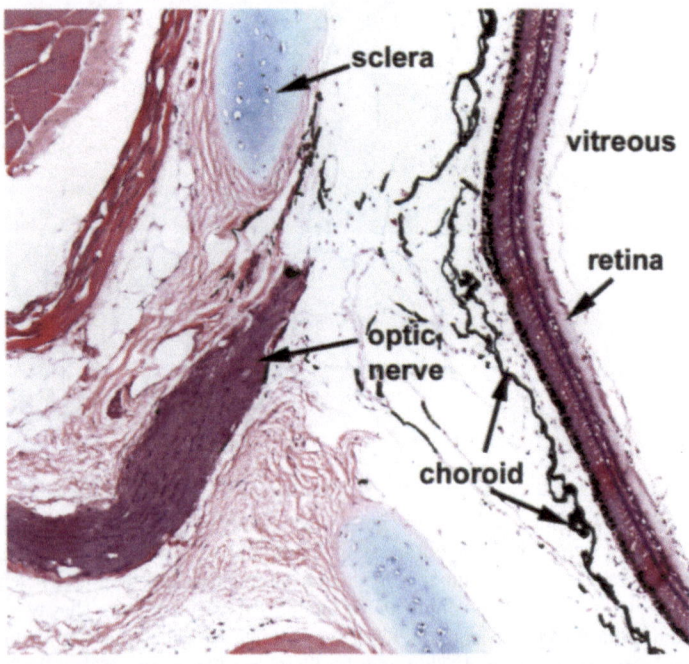

sclera

vitreous

retina

optic nerve

choroid

Figure 67. Image of the retina and optic nerve. The optic nerve leaves the retina at the caudal region of the eye. The optic nerve contains the axons of the ganglion cells of the retina. These axons project to various nuclei of the brain after passing through the optic chiasm to travel along the optic tracts. There is an open area of the hyaline cartilage of the sclera where the optic nerve projects from the eye.

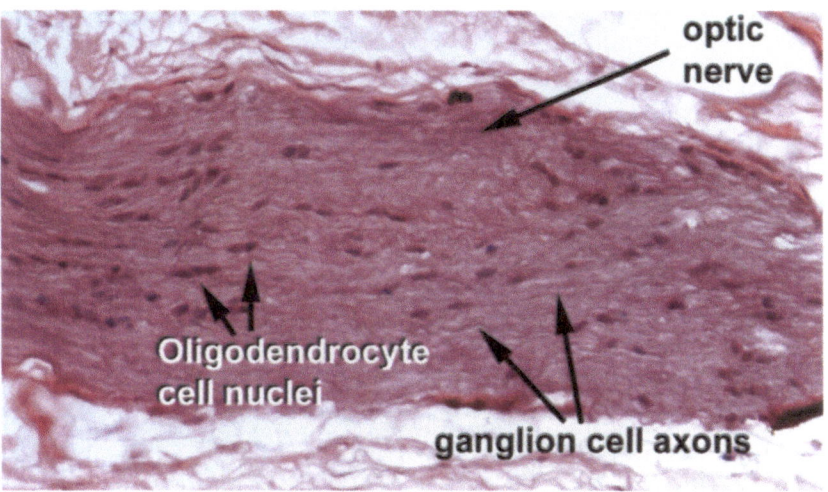

Figure 68. High magnification of the optic nerve. The optic nerve consists of myelinated ganglion cell axons projecting from the neural retina to the optic tectum and visual areas in the brain. Nuclei of myelin-producing oligodendrocytes are observed throughout the nerve. The nerve is surrounded by a connective tissue epineurium.

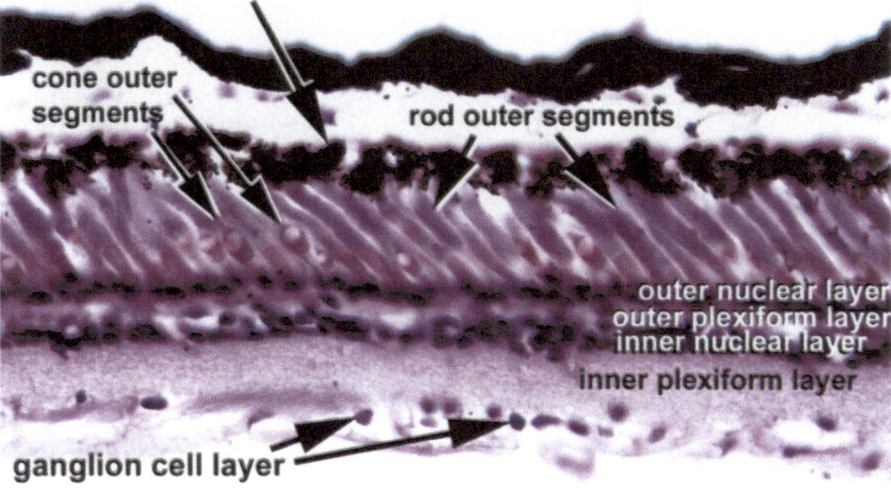

Figure 69. High magnification image of the retina. The retina is a layered structure. The photoreceptors respond to photons of light that pass through the retina, and the neural signal is transmitted through the retinal layers to the ganglion cells. There are both rod and cone photoreceptors in the *Xenopus* retina, and the distal portions of the rod outer segments are intimately associated with the apical microvilli of the pigment epithelium.

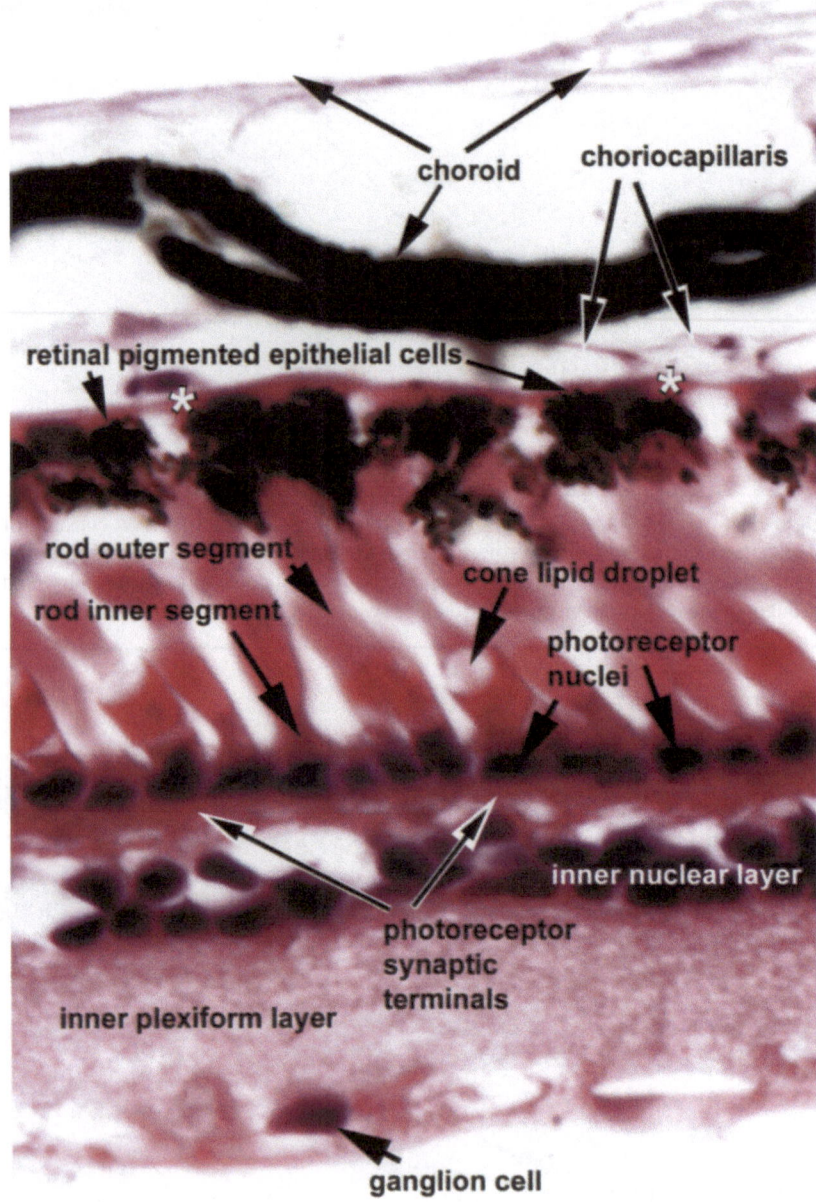

Figure 70. High magnification image of the *Xenopus laevis* retina. The outer choroid layer is a pigmented vascular layer. The choriocapillaris is a capillary layer in the choroid and is in direct contact with the basal surface of the retinal pigmented epithelium (RPE). It supplies the photoreceptors with oxygen and nutrients. The fused basement membrane of the choriocapillaris and the RPE is termed Bruch's membrane (*). The RPE cells phagocytize distal photoreceptor outer segments that are shed under a cyclic rhythm.

Index